Reinventing the Future

✦ ✦ ✦

Also by Thomas A. Bass

The Eudaemonic Pie
Camping with the Prince and Other Tales of Science in Africa

Reinventing the Future

✦ ✦ ✦

Conversations with the World's Leading Scientists

Thomas A. Bass

Addison-Wesley Publishing Company

Reading, Massachusetts Menlo Park, California New York
Don Mills, Ontario Wokingham, England Amsterdam
Bonn Sydney Singapore Tokyo Madrid San Juan Paris
Seoul Milan Mexico City Taipei

Library of Congress Cataloging-in-Publication Data

Bass, Thomas A.
Reinventing the future : conversations with the world's leading scientists / Thomas A. Bass.
p. cm.
Includes bibliographical references.
ISBN 0-201-62642-X
1. Scientists—Interviews. 2. Research—History. I. Title.
Q141.B32 1994
574'.092'2—dc20 93-14576
CIP

Jacket design by Lynne Reed
Text design by Joyce Weston
Set in 10.5-point Sabon by DEKR
This book is printed on acid-free paper.

1 2 3 4 5 6 7 8 9 10-MA-96959493
First printing, January 1994

Contents

✦ ✦ ✦

For my father

Ideas, one has so very few of them
in a lifetime!

—ALBERT EINSTEIN

Reinventing the Future

✦ ✦ ✦

Introduction

✦ ✦ ✦

SCIENCE is the dominant metaphor of the twentieth century. Today we all speak the language of probability. Science is the knowledge in which we place our faith, the solution to our problems, the way out, the way up. It is the only form of worship known to work.

At the same time science is a business as bureaucratized as any other. The corporate conduct of science tends to bleach out the drama. Acts have no agents, events no locus, time no turning points. "Science is the science of nobody," says Etienne-Emile Baulieu. "It is the science of robots, computers, and men in white coats." None of the scientists in this book is wearing a white coat, and two of them are women.

Hoping they could draw me a road map into the twenty-first century, I set out to interview the best scientists in their fields. It is probably no accident that I also found the most political. "I have never believed our way of thinking about science is separate from our way of thinking about life," says Mary-Claire King. "Whether we realize it or not, we are all political animals."

Each of the careers described here underwent a similar turning point. There was a shock of recognition, a crystalline moment of knowing that old ways of looking at the world no longer pertained. It is easy after the fact to talk about scientific revolutions, difficult to lead them. In spite of their Nobel Prizes and other accolades, the scientists in this book see themselves as outsiders and rebels. They would agree with Luc Montagnier when he says he has more enemies than most people.

Montagnier's story provides a good example. The accountant's son finishes medical school and leaves France for want of a job. The "outsider" goes to England, where he "destroys" two research laboratories by making their key discoveries. He returns to France and isolates the AIDS virus, which makes him internationally renowned and locally suspect. "All of a sudden a man in a white coat was mixed up with sex and drugs," says his colleague, Etienne-Emile Baulieu, whose abortion pill has produced its own collision among sex, drugs, and science. Montagnier confronts new enemies across the sea. He sues Robert Gallo and Gallo's employer, the United States government, for stealing his patent rights. Still raging are the scientific debate, the AIDS epidemic, and Montagnier's fierce desire to find a cure.

How they plotted their revolutions and at what cost they won them are described by ten other scientists: Sarah Hrdy, an expert on sexual strategies, both primate and human; James Black, inventor of the world's first billion-dollar drugs; Thomas Adeoye Lambo, Africa's first psychiatrist; Etienne-Emile Baulieu, developer of RU 486, the abortion pill; Richard Dawkins, proponent of the selfish gene's-eye view of evolution; Farouk El-Baz, extraterrestrial geologist now exploring ancient Egypt; Bert Sakmann, specialist on mind-body communication; Jonathan Mann, public health expert on global epidemics; Norman Packard, a physicist applying chaos theory to beating the stock market; and Mary-Claire King, the geneticist who discovered the gene responsible for inherited breast cancer and who now uses her genetic markers to recover children abducted during Argentina's Dirty War.

Is this a representative sample of scientific research at the end of the twentieth century? No. For one thing, it overrepresents women and Third World scientists in what is still a predominantly male, First World enterprise. If the people interviewed are not representative, are their disciplines? The answer may be yes, with several lacunae and a few extras. At the forefront of scientific research today is manipulation of the human genome; research in computers, artificial life, and the brain; medical chemistry; chaos theory and nonlinear dynamics; sociobiology; epidemiology; and the extension of our technology throughout this world and others.

Molecular biology, genetics, and chemistry are viewed here through the multiple perspectives of Luc Montagnier, James Black, Etienne-Emile Baulieu, and Mary-Claire King. Bert Sakmann takes us on a tour of modern research on the brain. Norman Packard and Richard Dawkins tackle the computer revolution and artificial life. Sociobiology, with its controversial application of evolutionary theory to the study of behavior, both primate and human, is the gauntlet thrown down by Sarah Hrdy and Richard Dawkins. Luc Montagnier and Jonathan Mann alert us to the global repercussions of AIDS and future pandemics. A geologist presents an example of technology reinventing a discipline as old as archaeology, and a psychiatrist reminds us that we have a lot to learn from other non-Western forms of science.

Scientists speak a variety of technical languages, which operate as a kind of shorthand, but none of these languages is used here, at least none that goes undefined. The point was to make these interviews as eloquent as the ideas they contain. I was looking for pure speech: charged with personal intonations and idiosyncrasies, yet stripped of qualifications, redundancies, jargon—all the linguistic noise that clutters even the best of our everyday discourse.

But how was I going to accomplish this feat without doing violence to people's ideas? The solution presented itself from the first discussion, with sociobiologist Sarah Hrdy, a talented author in her own right. Hrdy rose to the challenge and convinced me the same method should be employed for all these exchanges. I would engage the interviewee in writing the interview. I transcribed recorded conversations lasting up to four days, trimmed the manuscripts to a manageable size, and sent them off for comments.

"This is the stupidest thing I ever heard!" yelled a publisher who had commissioned one of these pieces for his magazine. "We never show advance copy to anyone." This magazine usually runs interviews with rock stars. Prone to altered states of consciousness and too much legal advice, rock stars may be unreliable editors. But this is not true of scientists, as I discovered with Hrdy and continued to rediscover with each successive exchange. Hrdy gave a personally revealing interview, but instead of X-ing out controversial statements, she *added* to them. She covered the manuscript with examples and anecdotes, and then pushed her argument even further in a long cover letter.

This is not to say these interviews were collected without difficulty. My darkest moment came with James Black. Our discussion nearly ended after the first question, when Black launched into a tirade against "public stripping" and made dark allusions to people trying to steal his "mental trick" and sell it. To keep the conversation going, I retreated to the high ground of analytical pharmacology, his specialty. But like the return of the repressed, Black's personality kept popping out, and all my questions eventually did get answered.

After spending the morning in his college office, I was invited to visit the James Black Foundation, which turned out to be a high-security laboratory newly built for Black by Johnson & Johnson. We walked past video cameras and opened locked doors guarded by code boxes to enter a series of sparkling white rooms filled with organ baths. These biochemical alembics held slivers of organic tissue whose response to various compounds was being monitored by computer. Never before having seen an organ bath, I asked some questions about how they work—too many questions, apparently.

I returned to New York to discover a message from Black accusing me of trying to steal his mental trick! I wrote to explain that I was incapable of industrial espionage; my questions were due to ignorance not intrigue. With no response, I sent another letter. Then I sent the transcript of our conversation. Then I phoned him twice a week for six months without getting past his secretary, until one day Black came on the line. "You win," he said. "I don't want to bankrupt you. Run the interview."

Nobody believes anymore in the myth of scientific objectivity. Science is a cultural artifact, a belief system lodged in history. Yes, experiments are verifiable and causes have effects, but the scientific enterprise is directed by value-laden assumptions. Change the assumptions, and you reinvent reality. Luc Montagnier distinguishes between "two kinds of scientists: the explorers who set out to discover new territories—either an island or an entire continent—and those who occupy these territories and build structures on them." The scientists in this book are explorers. They have changed the rules of the game—altered our perception of reality and the language used to describe it.

As for the arrangement of the interviews, I approached the problem like an eleven-course meal. What tastes complemented each other? Whose conversation sparkled next to his or her neighbors'? Dating the interviews is less than obvious. Some conversations are still ongoing. Others involved written exchanges lasting months or years after the first encounter. Nonetheless, I list the year and place of initial contact: Sarah Hrdy, Davis, California (1987), Luc Montagnier, Paris (1988), Richard Dawkins, Oxford (1988), James Black, London (1989), Farouk El-Baz, Boston (1989), Jonathan Mann, Geneva (1990), Etienne-Emile Baulieu, Paris (1990), Thomas Adeoye Lambo, Washington (1990), Norman Packard, Turin and Milan (1990), Bert Sakmann, Heidelberg (1992), and Mary-Claire King, Berkeley (1992).

Regarding my own participation in these exchanges, whatever labor may have gone into arranging them, I keep my presence in the text to a minimum. I leave to the reader the task of figuring out which answers required extra coaxing. I learned a tremendous amount from these encounters, which is one reason why I thought of collecting them in a book. On many occasions during these discussions I was surprised by new ideas and ways of conceiving the world. In chaos theory, as Norman Packard explains it, information equals surprise, and the more information you have, the more surprised you are. This has been my overarching principle in shaping this book: to maximize surprise.

Sarah Hrdy

✦ ✦ ✦

WHAT do women want? According to Darwin, they want the one best male. But after years of observing animal behavior, where females can be found soliciting male after male, behavioral biologist Sarah Hrdy has posited new rules for sex and the mating game. The second myth she has laid to rest is that of the coy female biologically fated to monogamy and sexual passivity. *The Langurs of Abu: Female and Male Strategies of Reproduction,* Hrdy's classic study of the sacred hanuman monkeys of India, depicts female primates as masterful sexual strategists.

Key theories in sociobiology and animal behavior—dealing with female sexuality, competition, orgasm, and infanticide—have been framed by Hrdy in her four published books and numerous articles. An eloquent writer with hidden aspirations as a novelist, she is not afraid to feature herself as a character in her own research. But the fame she has won for her revolutionary ideas has carried a stiff price. "Sometimes I have the uncomfortable sensation that I am being set up by the media as a spokeswoman for the naturalness of promiscuity, which, if you are talking about humans, I most definitely am not," she says vehemently.

Born in 1946, Hrdy wrote her first book, *The Black-man of Zinacantan: A Central American Legend,* as her undergraduate thesis at Radcliffe College. The book analyzes the Mayan myth of the bat-demon H'ik'al, a fiendish figure who takes vengeance on misbehaving women. For someone who later jumped the track from cultural anthropology to primate studies, which she pursued as a graduate student at Harvard, Hrdy's book looks like a precocious sport, the false start to a career not taken. But then one realizes that the social control of female sexuality is a crucial issue in all her work.

When the twenty-five year-old Hrdy went off to Rajasthan to study langurs on sacred Mount Abu (where Shiva's big toe is said to be enshrined), two facts about these graceful, black-faced monkeys struck her immediately. One was the promiscuity of the females, who habitually drifted away from their harem masters to "steal" copulations with roving males. The second had to do with the violence of these males. After they succeeded in taking over a harem, they were likely to kill all the unweaned infants, thereby bringing the females back into heat and readying them to produce new offspring that they themselves would father.

Hrdy, in a brilliant stroke, put these two observations together. What defense could the physically weaker females offer against males acting in their own biological self-interest? That's when she realized the importance of promiscuity: Female langurs were doing everything they could to confuse the issue of paternity, because no male would risk killing his own offspring. With every male a potential enemy, the females adopted fraternization as the best strategy for ensuring their infants' survival.

When first published in 1977, *The Langurs of Abu* was praised as an example of how field research could be used to generate novel theory. But Hrdy thought her critics had missed the point. Although her study evenhandedly detailed male and female sexual strategies, most reviewers focused only on the male half of the equation. Six years later Hrdy forced the issue by publishing *The Woman That Never Evolved.* Here females, both primate and human, take center stage. A blend of sociobiology, primatology, and feminism, the book scrutinizes primate sexuality and its relation to our own. Lest her critics misread her again, Hrdy wrote a disclaimer into her title. What makes us different from other animals, she says, is our ability to change our lives. If we value freedom and equality for both sexes, we can choose to create a modern woman who might never have evolved through nature.

Hrdy in person is a dynamo, spinning out ideas a mile a minute. Despite years in India, she still speaks with the drawl of a native Texan. During a daylong conversation that began at the University of California at Davis, where Hrdy teaches anthropology, we made several trips home to nurse her infant son, the youngest of three children. Later we drove out of town to the Vacca hills, where Hrdy and her husband, a medical specialist in infectious diseases, were building a house. She wanted to tell the bulldozer operator digging the reservoir for their orchard to make the sides *irregular.* "It will make a better habitat for wildlife," she said. "These hills look a lot like India, only the monkeys are missing."

✦ ✦ ✦

Is promiscuity in monkeys relevant to humans?

It seems an odd time to be asking that question. If we accept the argument given by Harvard's Max Essex that AIDS first evolved among monkeys, their promiscuity is going to be affecting human behavior for many years to come. From my point of view it seems no accident that AIDS evolved in vervet monkeys, because of the promiscuous breeding system that characterizes vervets—males and females having numerous sexual partners. That system created an ideal niche for the virus, with a ready mode of transmission. AIDS-like viruses have probably existed among monkeys for a long time. Virus and host must have come to a modus vivendi. But human promiscuity on the scale that it's practiced in urban Africa and among some urban homosexuals is a fairly recent thing.

Were you always interested in studying animal behavior?

Getting into primatology was an accident. My early research had been on the structural analysis of myths. I worked with Mayan-speaking people in Guatemala and Honduras for my undergraduate honors thesis at Radcliffe. I finished school and thought, *This is a lot of fun, but I want to do something relevant to the world.* One thing I'd done in Central America was teach hygiene courses for adults; so I decided to go to Stanford and learn to make films for people in underdeveloped countries. But filmmaking has got to be the weakest program Stanford ever invented. The guys were on the phone to Hollywood all the time, and when they weren't on the phone to Hollywood, they didn't feel important.

While I was at Stanford, Paul Ehrlich was giving a marvelous course on population biology. I cut classes in communications school to audit it. Ehrlich kept talking about overpopulation, and I said to myself, *If you really want to do something useful, why don't you study the effects of crowding on behavior?* It was an incredibly naive dream. No one has ever done a successful study of crowding on humans. I had taken a course on primate behavior at Radcliffe, and although I didn't learn anything because I read the books the night before the exam, out of that haze I recalled that Yukimaru Sugiyama, a Japanese primatologist, had reported that males of a monkey species called hanuman langurs were thought to kill their infants. And it happened only in areas where they were crowded. So I decided to go off to India to study his hypothesis—that crowding produced infanticide.

What made you reject his theory?

I chose to study a very high density population of langurs at Mount Abu, but I soon realized the males there were actually quite tolerant of infants. You would see them swinging on a male's tail or jumping up and down on him like a trampoline. Aggression against infants came only from males entering the breeding system from outside. Langurs typically live in troops of one adult male and several females. But bands of up to sixty or more males circulate around the outskirts of these troops. Attacks on infants come from outsider males trying to take over a troop.

So I realized the "pathology" hypothesis was wrong. Sexual selection provided a better explanation for what was happening. By eliminating unweaned infants and thereby inducing the females to ovulate, the new head of the troop was furthering his own genetic interests. He was compressing the females' reproductive careers into the brief span of time he could expect to survive as head of a troop, about twenty-seven months.

This is still looking at reproductive strategy from the male point of view. When did you focus on the females?

Watching these animals day after day, I began to identify with the females' problem. They were producing offspring, and then every twenty-seven months, on average, a new male would come along and kill the infants. Then the females would breed with the killer. Why would a female put up with this?

I saw that a female couldn't refuse to breed with an infanticidal male. She had to breed, because if she didn't, she'd produce fewer offspring than the competing females that did breed. Female langurs couldn't achieve a level of solidarity sufficient to stamp out this detrimental trait. Essentially, what allowed infanticide to persist was competition among females for breeding success.

It was a revelation to me when I realized that's what was keeping the vicious cycle in motion. I then asked: What other options are open to the female to keep a male from killing her infant? Some very peculiar things were going on! Females were mating not only with the infanticidal males, but with many or all of the males that accompanied him when he took over the troop. Even pregnant females were out there mating. Then it occurred to me that if they couldn't win by force, perhaps the most feasible reproductive strategy for a female would be to confuse the issue of paternity. Any male that kills his own infant is going to be rapidly selected against, so no male could afford to kill an infant that might possibly be his.

What happened when you published your findings?

I was attacked by some of the most eminent anthropologists in the country. They said my evidence was inadequate and that the animals I was studying must be crazy. Their concept of primate interaction was the old Radcliffe-Brownian model [developed by British anthropologist Alfred Radcliffe-Brown] of social organization. There, every individual in the group has a role to play that promotes the integration and prosperity of the group. Within that paradigm, how can an individual do something counter to the group interest? And what is more counter than eliminating infants? Anthropologists couldn't believe it was happening, which brings us back to the idea of scientific paradigms and the assumptions we start out with. Few of us are aware of how powerfully these assumptions shape the research questions we ask and the observations we make.

Your critics believed your earlier hypothesis, that infanticide is a pathological response to overcrowding. But you showed that infanticide occurs even when there is no overcrowding.

That's abundantly clear now. It wasn't clear then. The langur case was sexually selective infanticide: Males were killing the offspring of females who had bred with other males. These males were unrelated to the infants they killed. There are five different kinds of infanticide, each with its own functional explanation. Evidence from a wide variety of researchers is pulled together in a book Glenn Hausfater and I edited in 1984. The book was sufficiently convincing that the controversy is now over. Very few serious biologists today are not convinced that infanticide can be adaptive behavior.

What are the five types?

The first would be exploitation of an infant as a resource. That's cannibalism. Fish and insects provide the classic instances. The second type is competition for resources. By eliminating the infant, the killer increases availability of resources for himself and kin. The work of Cornell's Paul Sherman on Belding's ground squirrels would be an example. Females eliminate another mother's offspring, who abandons her perilous burrow, thereby making it available to the killer for her own brood. The third is Darwin's sexual selection—that is, competition between males for access to females, as among the langurs. Occasionally it can be due to competition among females for males. This happens in some bird species, like jicana, where the males tend the babies, and the females move from male nest site to male nest site laying eggs and occasionally destroying the eggs of other females. In either case, the gain to the killer is additional opportunity to produce offspring.

The fourth type is parental manipulation, and humans are the parental-manipulator infanticides par excellence. Usually in primitive societies infanticide occurs right after birth, and it's typically done with regret. A classic case in a hunter-gatherer society would be when the birth space is too short and a mother knows her four-year-old will die because there aren't enough resources to invest in both offspring at once. Number five is social pathology, which may involve total disruption and breakdown of the social system. Imagine the case of a captive mouse disturbed as she's giving birth, with the result that she eats her offspring. This is probably a haywire incident. There's no gain of any kind for the killer.

How do you feel about vanquishing your critics?

I like to play poker, especially a game called anaconda. If you have a low hand and the other players are high, you know you can afford to put up your chips. In the middle of the argument, when everybody was telling me I wasn't seeing what I was seeing, I felt like I had a pat low; I knew I was right. I'm normally fairly cautious in proposing ideas, but I didn't feel timid about my theories on infanticide.

Isn't this a Hobbesian world you're describing: nasty, brutish, and short?

Yes, and I wouldn't recommend to anyone that they use langurs as a template for living. In the last sentence of *The Langurs of Abu* I wrote, "For generations langur females have possessed the means to control their own destinies, [but] caught in an evolutionary trap, they have never been able to use them." That evolutionary trap, of course, is competition among females. My next book, *The Woman That Never Evolved,* was really an attempt to understand this female–female competition.

Anthropologist Donald Symons sees prostitution as the main female–female competitive strategy, with females soliciting favors in exchange for service to males. Does your theory differ from his?

In one of several scenarios he presents, Don suggests that females extended their periods of sexual receptivity in order to solicit favors from males. He imagines a chimplike ancestor. Male chimps do much of the hunting and share their prey with females. Don's argument is that a female in estrus could get more meat from the male when he's passing out tidbits. It's a plausible hypothesis, but there are plenty of other routes by which prolonged receptivity could have evolved. My own idea is that females, by extending their periods of receptivity and concealing ovulation, could mate with the resident male and at the same time mate opportunistically with other males as well. One female counterstrategy to infanticide is continuous sexual receptivity.

Isn't your idea really a variation on Symons's prostitution hypothesis—except in your case the prostitutes are doing it to keep their children alive?

I've thought about this. Who is the object, and who is the subject? In the prostitution hypothesis, the male (subject) is offering the female (object) something in order to get what he wants. In the manipulation hypothesis, the female (subject) has a good deal more control over the situation. See the distinction? But even if the hand that rocks the cradle rules the world, basically, from a sociobiological point of view, it's bad luck to be born either sex. The kind of civilized world I try to construct for myself—in which I don't compete with other females and try to maintain a high degree of trust and loyalty within my family while raising my children to respect the rights of others and live by egalitarian standards—that's a world that never evolved. It's an artificially constructed world. And living that way is a much more heroic endeavor than simply living as our ancestors did.

Symons says the female orgasm is nonadaptive; the clitoris is a vestigial organ serving no reproductive purpose. Your own view seems subtly different—that the female orgasm doesn't do what it was once designed to do.

We both agree that the female orgasm is currently not adaptive. If it were adaptive, the morphology [structure] of the female genitalia would be quite different. If the female orgasm evolved, as some people argue, to cement the pair bond and enhance marital relations, you would expect it to be as reliable as the male orgasm, and it isn't. A male has an orgasm every time he ejaculates. The female, in the majority of cases, does not have an orgasm in response to copulation alone. If the female orgasm evolved in order to cement the pair bond, one has to have a dismal view of evolution. Natural selection is often less than perfect, but this is really substandard! The idea that somehow the orgasm makes it easier for the female to be satisfied by one male is nonsense. If we have learned anything about female sexuality, it is that only about thirty percent of women have orgasms from intercourse alone. Furthermore, there is a disconcerting mismatch between a female who is capable of multiple, sequential orgasms paired with a single male who is typically capable of only one orgasm per copulatory bout.

What do you think about Symons's contention that the clitoris is a male trace left on the female body and that the female's capacity for multiple orgasms is compensation for failed ejaculation?

Symons, and recently Stephen Jay Gould in *Natural History*, stress that developmentally the clitoris is nothing more than a vestigial homologue to the male's glans penis. Like a male nipple, it is present incidentally in one sex because it is essential to the reproductive system of the other. The simplicity of this argument is one of its strengths. But if the clitoris is nothing more than a homologue of the penis, we would expect it to parallel evolutionary changes in the penis, and it doesn't always. What bothers me is that Gould, especially, claims the female orgasm is not adaptive, period. I say it could well have been adaptive *at one time*. Let's take a look at the evidence.

I agree that it is certainly not adaptive for human females to run around behaving promiscuously. They have a lot more to lose than gain in our society and probably in societies for many thousands of years preceding ours. One of the most mortifying reviews I got for *The Woman That Never Evolved* said I believed human females are sexually insatiable and constantly going around looking for sexual partners. Anyone reading that review must have thought I was nuts, as I sure would have been had I said that. The reviewer had taken out of context something I said about female Barbary macaques, monkeys who in fact appear to be sexually insatiable when they first come into heat. When she comes into estrus, a Barbary macaque solicits several males an hour. She's switching consorts every five minutes. I can assure you that in species like macaques, savanna baboons, or chimpanzees, females are doing anything but mating with just one male!

But the question is: How do you explain the contrast between a macaque and the more sedate and self-conscious sexuality of the human female? I argue that while the female orgasm is not currently adaptive, it probably once was. It provided the motivational underpinnings for females to mate with a range of partners, thereby confusing the issue of paternity. This increases the likelihood that the female can extract investment—like food and protection for her young—from a number of different males, or at least ensure their tolerance for her offspring. Remember, infanticide by males who couldn't possibly be the father is a very widespread threat for primates and has been for a long time.

What would Darwin have said about your theory?

Darwin was a classic Victorian in his attitudes towards women. As brilliant as he was, and as heroic, he was clearly a sexist who educated his sons differently than his daughters. He couldn't imagine that females would typically solicit more than one male. His theory of sexual selection posits that males compete among themselves for access to females and that the females choose the best male. It's a great theory, and we've gotten a lot of mileage out of it in behavioral biology. But its central assumption has led to all kinds of false premises.

My view of the world is very different from the Darwinian tradition of thinking in terms of the single best male. How could the female orgasm be adaptive if all females want to do is mate with one male? That's a very male assumption, and I think it's the way most men think. They're convinced that women are out there looking for the best male, and they're trying hard to be that best male. But is that really what's going on? My point of view is shaped from many years of observing nonhuman primate females soliciting male after male, and then asking myself, "What on earth are they doing?"

I've often wondered what would have happened if Darwin had lived not in the nineteenth century but in the seventeenth, when females were seen as sexually assertive, bawdy creatures, not coy and passive. If the theory of evolution had been published then, our idea of sexual selection would be very different, and much closer to the kind of approach I take today.

If it's not adaptive now, why hasn't the female orgasm been lost?

One possibility is that it's on the way out. Maybe that's why it can be so erratic. The things I'm suggesting are testable. For example, if natural selection has been operating on the clitoris, you would expect it to be more pronounced in primates that have multimale breeding systems—where females mate with a number of males. Is this a fact or not? What sort of anatomical variation in the clitoris exists across species? Chimpanzees and orangutans have penises that are relatively and absolutely smaller than human penises. Yet the clitoris appears to have evolved in the other direction: It's much larger in chimps and orangutans than in humans.

I doubt we will ever be so lucky as to find the reproductive equivalent of the hominid fossil footprints discovered by Mary Leakey at Laetoli, so we may never know about the genitalia of our human ancestors. But we can systematically look at comparative evidence from other primates, although no one has done it yet. And if they did, there would be such a stigma attached to their research that they might never find a job. The only person who could safely study this taboo realm of female sexuality might be an emeritus woman professor of anatomy.

What evidence is there for orgasm in primates?

When Canadian anthropologist Frances Burton first proposed the idea in the late Sixties, she was scared to death. She didn't have tenure, and she feared it might hurt her reputation. But the idea has occurred to anyone who has watched animals mate. You see a female's intense response and wonder. Here, for example, is a photo of a female clutch response, with the female looking back at the male and grunting. This look precedes what some of us think may be indicative of female orgasm in primates. Orgasm itself is signalled by the round-mouthed "O" face. Here's a picture of it in stump-tailed macaques. See how the lips are pursed together in an oval shape? Now look at this: a picture of Gian Bernini's sculpture of Saint Theresa in ecstasy. The same perfect "O" face! The sociobiology of religious ecstasy—that should add to the controversy.

Scientists have argued that sexual encounters between animals are too brief to result in orgasm.

The old argument that orgasm can't occur in nature is clearly wrong once you realize that females solicit many different males. The stimulation needed to produce orgasm in females is cumulative. It requires more stimulation than is typical of a single bout of mating in the wild. But the opportunity is there for it to occur—provided the female mates with a series of males within a short period of time.

***You made a movie called* Stolen Copulations.**

Yes, together with my husband, Dan, and the filmmaker John Melville Bishop.

What's the plot?

It starts with a band of male langurs mounting each other and getting more and more excited. This is what I call "revving up"—generating a level of excitement and solidarity before the males go about the dangerous business of "stealing" copulations. Homosexuality in langurs is part of a heterosexual strategy. I'm not ruling out the possibility of exclusively homosexual behavior in primates. But we don't have data on lifetime sexual preferences for any animals, and that's what you need before you can say homosexuality is a stage in another strategy, or a separate option.

What do the males do after revving up?

They approach a troop where the dominant male—whom we called Harelip, because his upper lip didn't quite cover his teeth—has been successful enough to hang on for nine years. Some of the females in the troop are his daughters. They had the highest incidence of extra-troop copulations, which may have been an incest avoidance mechanism. Anyway, the females are wildly soliciting these males. Harelip has gone off to forage. Then he comes racing back and chases these guys off. He's not interested in mating with the females; they're not interested in mating with him. They're interested in the outsiders, but he keeps chasing them off.

Do they succeed in stealing copulations?

Sure. But some of the males were quite young and inexperienced. Here's a picture of one of them mounting a female. Other males in the group are harassing him, and he's not correctly mounted. He was just learning to mate.

You sometimes use portraits of your children to illustrate primate facial expressions. How relevant is it to extrapolate from primate to human behavior?

The grimace of nonhuman primates relates to the human smile. So, too, do the openmouthed expression and the "O" face during orgasm. At some level I'm convinced that primate motivations are similar to ours. Something like sexual jealousy must be an old emotion. It's a human universal, and animals know it, too.

How does a sociobiologist talk about jealousy? It seems to run counter to your ideas about multimale solicitation.

Certainly if *any* human behavior has an evolutionary basis, jealousy is going to be it. Even in multimale breeding systems, the male does not like to see a female breeding with other males. In chimpanzees, several males mate with a female. But a male not currently breeding with her is, first, sexually excited, and second, upset at seeing the female with other males, even though he may tolerate it. "Upset" is an anthropomorphic term, but that's the only way to describe it. Primate males do not enjoy watching other males copulate with females they're interested in.

Is the absence of rape among primates one reason you say in* The Woman That Never Evolved *that human females are worse off than other female animals?

That wasn't what I was thinking of. I agree with the early social philosophers like Engels that men in patriarchal societies try to control the sexuality of women, a sexuality perceived as dangerous. This happens when you start to get complex, stratified societies based on the ownership of property. You then become exceedingly involved in the inheritance of property, and you worry about adultery because you want to avoid the confusion of family lines. The whole business of paternity takes on new dimensions when there's property and language and gossip. The sexual activity of women is of consuming interest in all cultures. We watch soap operas on TV, while in other societies people gossip endlessly. And what are they gossiping about? Women's sexuality. What do they do about it? Curtail women's freedom of movement and with it so many other opportunities.

In what other ways are human females worse off?

They are much more likely to be cloistered than other female primates. Female baboons may be beaten up by males, and you have beatings occurring among chimps and gorillas. Nevertheless, females in these species have a lot more freedom of movement than humans. In agricultural societies, where property is inherited through the male line, the threat of female sexuality causes a lot of anxiety, and people try to control females. There are several ways to do this: One is to indoctrinate women with puritanical views. More extreme and crueler forms include female circumcision, where you excise the portions of the female anatomy that might permit some kind of sexual pleasure.

One of the things that worries me about teaching this stuff is how excited my students get by these "what if" evolutionary scenarios. They are intrigued by the entire field of reproductive strategies; they can speculate endlessly about mate choice. People's interests track their life phase. Which brings up the more serious issue of whether this stuff should be taught at all. I question whether sociobiology should be taught at the high school level, or even the undergraduate level, because it's very threatening to students still in the process of shaping their own priorities. They're continually wondering if they're missing out on something.

Why is it threatening?

The whole message of sociobiology is oriented toward the success of the individual. It's Machiavellian, and unless a student has a moral framework already in place, we could be producing social monsters by teaching this. It really fits in very nicely with the yuppie "me first" ethos, which I prefer not to encourage. So I think it's important to have a certain level of maturity, with values already settled, before this kind of stuff is taught. I would hate to see sociobiology become a substitute for morality. It's a nasty way to live.

The idea that females are coy and subservient used to be anthropological dogma.

One big influence on my own work was the obvious bias of many of my predecessors in anthropology, who looked only at the males. They counted up the number of copulations a male had and called that his reproductive success. Jostling and alliance forming among males composed the social organization. Reacting to that, I focused instead on females. I probably overstressed the importance of females in primate social organization. So now it's time to look at the interrelationship, at how male reproductive strategies evolve in response to female reproductive strategies, and back and forth. It's time for a synthesis by a younger generation, less polarized than my own. One thing's for sure, the days are gone when females are looked at as passive objects in a landscape defined by active male protagonists. The amazing thing is how quickly it happened. Professors famous for their androcentric bias are now rushing to organize symposia on female strategies.

How has the new female perspective changed the field?

Take a look, for example, at Don Symons's book on *The Evolution of Human Sexuality*. It's a fine book, but it isn't about the evolution of human sexuality. It's about the evolution of male sexuality, and of one kind of male sexuality at that—what I call the "promiscuous Walter" male. Do you remember Walter, the hero of *My Secret Life,* the Victorian novel that chronicles the sexual experiences of an upper-class man who goes from chambermaid to chambermaid? I'm convinced, although I can't prove it with any scientific data, that male personalities exist along a continuum of male types. You have some males who are incapable of long-term relationships and are not interested in having and rearing children. One of the fascinating things about Don's book is that it barely mentions offspring, or children, or the factors that contribute to their survival. It's all about achieving copulations. I'm much more interested in reproduction as the production of children who survive than I am in reproduction as a sexual act—a moment in time.

You have said that competition among women, and their lack of solidarity, prevent them from assuming power.

Women make up more than half the population. So why have they traditionally been politically powerless? It's partly because they identify with their families, husbands, and children, and not with other women. We are crippled by the feelings of competitiveness and jealousy we feel for each other. Put women in a group that might be competitive, and they will compete not with the men in the group but with each other. Then you realize you can decide to run your life differently. I have made a point of being supportive to women colleagues and suppressing my competitive feelings, because I'm aware of the unpleasant alternatives and also because I remember with gratitude the support I received from older women colleagues early in my career. There were no women professors at Harvard when I graduated from Radcliffe in 1969. I mean zero, but as I started to look around, I found these wonderful women mentors in my field—like ecologist Alison Jolly and anthropologist Jane Lancaster.

Has there been a "feminization" of primatology?

Women like Jane Goodall, Biruté Galdikas, and Dian Fossey have spent years in the field without permanent positions, while the men were back home tending their power bases. Women distinguished themselves in this field because they were willing to do something that many men were not. In the process these women were shafted, a fact that gets eclipsed by their fame. Normal sources of funding for scientific field work, like the National Science Foundation, were not available to them.

They had to take to fund raising on the stump. When Fossey came back to the United States she had no home or permanent position, and her site in the Virunga mountains was being wrested from her. She was living from hand to mouth, taking teaching jobs where she could. Her health was failing. When she returned to Africa I had the feeling she was going back to die. It would have been nice if some of the flimflam after her death could have been changed into support during her lifetime.

It may also be true that in the particular area of primatology that has attracted so much public attention, women are better at studying behavior than men. If you can't recognize individual primates and follow them over the long term, you're not going to learn that much about them. I stayed out under the Indian sun watching hanuman langurs for hours at a time because I wanted to know what was going to happen next in the soap opera. Some men are able to memorize the traits that distinguish one primate from another, but by and large, women have done it better.

Does their natural empathy make women better at studying animal behavior?

A close friend of mine, a primatologist who studies baboons, once said to me, "I identify more with female baboons than I do with males of my own species." I went out to India to do a traditional study, in the Harvard mode, of male reproductive strategies, because that's what infanticide is partly about. But suddenly my research went through a dramatic shift in perspective. I started to identify with the hazards confronted by the female langur. It changed the way I thought not only about females, but also males.

You have transformed sociobiology by making female sexuality central to the theory in a way that it wasn't before.

A lot of people criticize sociobiology by talking about how sexist sociobiologists are. Sociobiology obviously has its share of sexists. Why should I deny that? I've suffered from it. As brilliant as Darwin was, he, too, was clearly a sexist. But that doesn't mean that evolutionary theory is wrong, or that sociobiology is not extremely useful for analyzing data about wild creatures and ourselves. Critics of sociobiology have confused the presence of sexists in the field with the approach itself.

You're certainly a primatologist and an anthropologist; so why do you call yourself a sociobiologist?

Because I think evolutionary theory is valuable for analyzing behavior—primate or human. I have certainly not lost any of my earlier interest in history and culture, but evolutionary theory makes me ask different questions about why beliefs get constructed the way they do. Look at my structural analysis of the H'ik'al myth in my first book, *The Black-man of Zinacantan.* H'ik'al is a little, super-sexed bat-demon, also represented as a black-man, who zips down and slices off people's heads and does other wild things. In the contemporary version, which is still told around camp fires in southern Mexico, he punishes women who have not properly fulfilled their sex roles. If the black-man catches a woman who has transgressed sexual norms, like going out at night without a chaperon, he rapes her and she gives birth, night after night until she finally swells up and dies. That's everyone's worst fantasy who has ever given birth by natural childbirth.

My book showed how history was important in shaping the contemporary view of the myth. But I didn't focus on the issue of why sex roles were defined that way in the first place. Why had the Mayans built up this whole mythology to keep women in line? Consciously or subconsciously, that's the question I was drawn to in my later research. I was totally innocent of feminist theory when I wrote this book. I had no inkling such things existed. But now it seems to me there are a lot of links between the structural analysis of Mayan myths and the kinds of things I wrote about in *The Woman That Never Evolved.* It just took me a long time to put it all together.

You've been a kind of Trojan horse in the profession, penetrating it with feminist ideas.

Only with great anguish was I a Trojan horse. I remember a seminar at my professor's house at Harvard when I heard Robin Fox talking about women being traded between groups of men. I sat there thinking to myself, *This is what it must be like to be a black person listening to a lecture on niggers and the Ku Klux Klan.* It was a horrible experience and unforgettable, but it taught me something about empathizing with people who are oppressed. It polarized me enough to write a book like *The Woman That Never Evolved.* There's a lot of theoretical tension in that book, because I myself was so polarized. It was only by listening to these views that I knew were biased and wrong that I came to understand what I really thought. I developed extra gray hairs and suffered from the experience, but I don't regret it.

Sociobiology has been accused of being biologically deterministic: We're programmed to further our individual genetic interests, and that's it.

An uneducated person who does not understand evolutionary theory might confuse natural selection with social Darwinism and eugenicist thinking. A pernicious right-wing group in France has adopted sociobiology while understanding almost nothing about evolutionary theory. But this is ignorance, and one really can't hold oneself responsible for other people's ignorance if the information has been made available to them.

What is your relation to the women's movement?

The women's movement is an experiment. There's never been anything like it. As heroic an endeavor as it is, I worry it may be fragile. If we mess up this opportunity, we may never again have the chance to create this kind of freedom for women. That's why I'm worried when I see women becoming overconfident and cavalier about the opportunities we've won. As I wrote in the afterword to *The Woman That Never Evolved,* "The female with 'equal rights' never evolved; she was invented and fought for consciously with intelligence, stubbornness, and courage."

How do you evaluate yourself as a feminist theoretician?

I don't believe there is anything called "feminist theory." If you take predictions generated by a hypothesis, and they test out positive, and you or other workers replicate these results, then it becomes a theory. But what feminist hypotheses have generated predictions sufficiently testable to be called a theory? I'm embarrassed by the phrase "feminist theory." It is not a theory. It's an approach, or perspective.

What are you working on now?

Some years ago I became interested in a very widespread phenomenon in nature: biased investment by parents in sons versus daughters. In some creatures and populations, sons are favored. In others, daughters are. The nutria—which is a kind of aquatic guinea pig with particularly lush and attractive fur—provides one example. The British zoologist Morris Gosling discovered that mothers pregnant with small litters reabsorb them if the litter is mostly male. In monkeys, mothers respond to one set of conditions by producing mostly daughters and to another by producing mostly sons. This is a new area of research and poorly understood. Together with Meredith Small, I have been studying factors that affect the production and survival of sons and daughters among several species of macaques.

As a kind of tangent on this research, Debra Judge and I have begun a more sociological and historical study of how parents divide their estates among male and female offspring. We are examining this question by focusing on three thousand wills made in California. We want to see how patterns have been changing over the last hundred years.

Are you writing another book?

Yes, on motherhood. My ability to do field primatology is compromised by having three children, so I figured I might as well turn a liability into an asset and write a book that pulls together a lot of the research I've been doing over the last five years. All of it revolves in one way or another around the issue of parental investment. The will research, for example, is really a study of how parents invest in their kids after death and the family strategies people devise to pass on resources.

Does one necessarily have a male or female perspective on these questions?

No. Once the initial leaps of imagination have been made, a man can see things as well as a woman. It's only when we have to revise the paradigms—make radical shifts in our intellectual assumptions—that I think being male or female, privileged or oppressed, white or black makes a difference. It's important in the initial empathizing required to create an imaginary world and move into it.

Your work has endowed females with cunning, duplicity, Machiavellian genius. . .

Ah, but the way you're saying these words! As if they're bad. Cunning and duplicity—these are the strategies of a creature who can't win a fight that depends on muscle mass. To criticize a creature for taking the only route available to it—given the limitations, the constraints, the oppression—is not fair. I worry that somebody could read misogyny, a hatred of women, into my work. What they should see instead is an effort to expand the range of attributes encompassed by a term like *female nature.* We've tried so hard to change stereotypes about females and say, Look, females are not just passive; they are also competitive and assertive creatures. But that is not to say they are *only* competitive, vicious, and cunning. They are in fact cooperative, nurturing, competitive, cunning, intelligent, and creative. Rather than reducing female nature to any single set of stereotypes, we want to talk about a whole range of potentialities.

Luc Montagnier

✦ ✦ ✦

"I'M regarded as a kind of magician. People phone me in the middle of the night begging for help. Out of necessity, I've had to build a wall around myself. My secretaries are instructed to steer people to clinics and doctors who work directly with patients. Even so, I find people waiting for me on my doorstep when I get to work in the morning," says Luc Montagnier, who discovered the virus that causes AIDS.

"AIDS is a disease of civilization," says the French biochemist, "a disease of the city." Yet he thinks the virus itself is old, so old it might well have appeared earlier in history to wipe out previous civilizations. Since then, the virus has been biding its time in isolated populations, masking its existence behind other fatal diseases—until the "cofactors" of modern life united to spur it into an epidemic. These cofactors weakening the immune system include sexual promiscuity, industrial pollution, drug use, and the mingling of people into a world culture. "The whole world is married to one another," says Montagnier. "The globalization of culture has globalized our germs."

Analysis of a sliver of tissue from the swollen lymph nodes of a homosexual man admitted to La Pitié Saltpêtrière Hospital in Paris led Montagnier, the director of the viral oncology unit at Paris's Institut Pasteur, to isolate the virus in early 1983. Quickly publishing his findings in the journal *Science* that May, he would spend the next year battling to convince his colleagues that acquired immune deficiency syndrome was caused by a virus—and by *this* virus in particular.

His major opponent in the United States was Robert Gallo of the National Cancer Institute in Bethesda, Maryland. Gallo held that a member of the leukemia-producing family of retroviruses that he had discovered caused AIDS. Montagnier agreed they were dealing with a retrovirus—a virus whose genetic material is made of RNA—but he kept telling Gallo that the AIDS virus and his leukemia viruses had opposite effects. Instead of causing cells to multiply uncontrollably, as did the leukemia viruses, the AIDS virus killed them.

At this stage in their careers, Montagnier and Gallo were friendly rivals in the same research. Twice in 1983 the French investigator sent samples of the new virus to his American colleague. Montagnier was dumbfounded when in April 1984, at a Washington press conference called by the United States secretary of health and human services, Gallo announced that *he* had discovered the AIDS virus, which he christened HTLV-3, the third in his series of T-cell leukemia viruses.

Scientists quickly confirmed that Gallo's virus was identical to Montagnier's. Gallo's claims to have worked independently of the French laboratory were further compromised when he "accidentally" published Montagnier's photographs of the virus in a *Science* article announcing his findings. Montagnier was further outraged when the United States patent for the AIDS blood test, which he had applied for in 1983, was awarded a year and a half later—to Robert Gallo. "I was furious," says Montagnier, who ended up suing Gallo and the United States government.

The two parties settled out of court in 1987—a settlement that has since unravelled, after further allegations of wrongdoing were levelled against Gallo. As announced at the White House by President Ronald Reagan and French prime minister Jacques Chirac, royalties on the AIDS blood test were to be split between the two countries. Most of the money would go to a foundation for AIDS research. Montagnier and Gallo would henceforth be known officially as "co-discoverers" of the virus. The dispute over what to call the virus was resolved by another independent commission, which settled on *human immunodeficiency virus*, or HIV. Two major strains of the AIDS virus are now recognized as HIV-1, which has infected millions of people in the United States and the rest of the world, and HIV-2, a West African strain Montagnier discovered in 1986.

The American press has branded Montagnier as patrician and aloof. This may appear to be the case when he's speaking English, but at the Institut Pasteur he is witty and outspoken. Among his staid colleagues Montagnier stands out as a scrappy figure freely acknowledging the motives behind his research, including ego and ambition. "I'm a self-made man," he says. "In my aggressivity, I'm really half American."

The grandson of French peasants and the only child of an accountant, Montagnier was born in the Loire valley in 1932. After studying natural sciences at the local university in Poitiers, he got his medical degree in Paris and then left France for four years to work in England and Scotland. At the viral research unit of the Medical Research Council (MRC) at Carshalton, south of London, and at MRC's Institute of Virology in Glasgow, Scotland, the young biochemist made the initial discoveries that allowed him to return home and work his way up to an appointment at the Institut Pasteur.

Montagnier's early research focused on cancer. His idée fixe was how certain cancers are caused by viruses. Montagnier was the first to show how single-stranded RNA viruses replicate by making a double helix. He then invented a technique for multiplying cancerous cells in agar, which is now standard laboratory procedure. Next he isolated the messenger RNA of interferon—proteins that stimulate immune cell activity. This led to its being cloned and commercialized. When Pasteur Vaccines asked him to look at a strange organism that might be contaminating their blood supply, this was supposed to be only one among many ongoing experiments, but Montagnier now spends all his time working on AIDS.

Walking past Louis Pasteur's statue and the house under which he is buried, one comes to the low brick building identified as the virus laboratory. Montagnier's office lies at the top of the stairs, where he presides over a long corridor of rooms overflowing with equipment and experiments in progress. His office is crammed with books and papers covering every available surface.

A short, energetic man in a rumpled suit and tie, Montagnier ascribes his small stature to malnutrition during World War II. The experience made him conscious of his health and the cofactors he thinks are crucial in depressing the immune system of AIDS patients. When he and I went to lunch at Le Récamier, a restaurant popular with writers and editors in the Saint-Germain quarter of Paris, Montagnier refused to eat on the terrace. He drives an air-conditioned car and lives outside the central city. Walking the streets of Paris—filled with diesel fumes and asbestos flaking off the brake shoes of cars—he considers a risky business. Once inside the restaurant, however, Montagnier ate with gusto and discoursed at length on the epidemic he thinks will soon be ended, if not by a vaccine, then by changes in human behavior.

While rushing to develop a vaccine, Montagnier also tests two hundred chemicals a week as potential drug therapies. He also researches the virus itself, in the hopes of outsmarting what he considers the world's most intelligent pathogen. This activity counters a lot of pressure for him to move out of the laboratory and into the public spotlight. "The last thing I want to be is an explorer with a statue, someone revered for a short time and then forgotten," he says. "I want to stay in the race."

✦ ✦ ✦

Why did you become a scientist?

My father's hobby was science. When the weather was good he went fishing. When it was bad, he tinkered in the basement making electrical batteries and things like that. My earliest memories are of my father doing his Sunday experiments. But when the war arrived in 1940, there was no more Sunday science. Originally I wanted to be an atomic physicist. But when I saw how many people were killed by the bomb dropped on Hiroshima in 1945, I said, "That's enough atomic energy for me." I was never good in math, so I couldn't have become a physicist anyway. By the time I was fourteen I had become an amateur chemist, making nitroglycerin in our basement laboratory. Then I turned to medicine and biology, thinking they might give me more concrete answers to my questions.

Was your family religious?

My family was Catholic—in France almost everyone is Catholic—but my father and grandfather were strongly anticlerical. So I also became a religious skeptic. I have no need of religion, at least as expressed by the great churches. But we face a contradiction. We're headed for a world in which ideology and religion no longer exist, at least in the developed countries, while at the same time we need to believe in something. This results in the urge to replace religion with science. But I don't want science to look like sorcery. This could lead to science being blamed as the cause of our woes. Fortunately the rumor didn't get very far, but people have even said that AIDS was created by scientists.

Was it?

That's absurd. To make a new virus, you need the kind of knowledge that molecular biology has attained only in the last ten years, but the AIDS virus is at least twenty-five years old. For someone to have created it, he would have had to have been way ahead of his time. But one could also imagine a scenario in which the virus was created accidentally in a laboratory. In Russia in the 1960s, and probably also in the United States, scientists injected monkeys with blood from leukemia patients to see if the disease was caused by a virus. In the process, they transmitted viruses from humans to monkeys. This leads one to wonder if they also transmitted viruses from monkeys to humans. Someone stuck himself with a needle and thereby became infected with a monkey virus. It's a possibility, although highly unlikely.

You were trained as a medical doctor. Why did you switch to research?

I knew right away I'd make a bad doctor. I'm not altruistic enough, and I don't like being around sick people all day. I wanted to do research related to human biology, but in France there was no training in this area outside of medical school. My parents were opposed to my doing something as risky as becoming a researcher. Their dream was for me to become the village doctor, make a lot of money, and live in a big house.

From the start, what drew you to cancer research?

There are two kinds of scientists: the explorers who set out to discover new territories—either an island or an entire continent—and those who occupy these territories and build structures on them. Both types of scientist are necessary, but clearly I'm one of the former, an explorer, an adventurer. What interests me are the great enigmas in biology, and one of these is cancer. On a more personal level, I saw my grandfather die of colon cancer. It lasted seven years, and he wasted away little by little. I was fifteen when he died, and I keenly remember how much he suffered.

Why study viruses?

There was no way at the time to attack the cancer problem directly. So I began working on viruses, which were easier to understand. This led to what I consider one of my major contributions—the discovery of how RNA viruses replicate. How can a virus containing only a single strand of RNA reproduce itself? Our goal was to discover the famous double helix, made this time out of two-stranded viral RNA rather than DNA. I was the first to observe this in England in 1963.

There is an element of luck in scientific research, but you have to put yourself in the way of being lucky. I had humble beginnings in a provincial school. Then at the Sorbonne I had the misfortune to fall on a patron who was scientifically mediocre. Because of this I had to go abroad to launch my scientific career. In leaving France I made a lot of enemies in the academic community. I'm a solitary figure, an individualist who doesn't like to follow the lead of other people.

I was lucky enough to fall on a good laboratory, which my discovery of RNA double helixes actually helped destroy. I completed its principal objective, so the lab was disbanded. After working for three years in London, at the age of thirty-one I was surrounded by a glow of success. But before returning to Paris, I went farther north to Glasgow, where my research on cancer began in earnest. In Scotland I began to look at animal oncogenes—cancer-producing viruses that I hoped to find in humans. Once again I played the role of the outsider who arrives to notice things that have been there all along but that no one has seen. Working in Glasgow with Ian MacPherson, and using an observation made by Kingsley Sanders in my London lab, I discovered how to grow cancerous cells in agar. This allowed me to return to France and apply my new technique to the search for cancer-producing viruses in humans. It's an idée fixe of mine that certain cancers are caused by viruses.

The second goal of my work at the time had to do with learning *how* these viruses replicate, and this naturally included the retroviruses. I failed in this search because I set out with the wrong model. We now know that retroviruses replicate by hijacking a cell's DNA, but I was looking only at their RNA. Nonetheless, I found some interesting things along the way, including double helixes of RNA that exist in normal cells having nothing to do with viruses.

I was also working on interferon, and in the early Seventies I and colleagues Jacqueline and Edward DeMaeyer were the first to isolate its messenger RNA. My experiment showed how you could extract the messenger RNA from the cells of a chicken and introduce them into a mouse. Because interferon is species specific, the mouse would start producing chicken interferon. Our findings eventually led to the cloning of messenger RNA. I would have liked to have been the first to do it, but, alas, I was not.

Why not?

It required a lot of money I didn't have. This happens a lot in France. We make discoveries that we have neither the means nor will to perfect; so other people benefit from them. Ever since I began working on AIDS, I've been haunted by this fear of being an explorer discovering things that other people develop. I don't want to repeat my experience with interferon. I will do everything possible to prevent this from happening again, which means I have to spend a lot of time fighting and make a lot of enemies.

The American press describes you as proud and ambitious to the point of arrogance. Are you?

It depends on the day. When you're climbing a mountain, the last thing you want to do is look behind you and say, "Oh my, it's too high, what am I doing up here?" Even if I keep my eyes fixed on the summit, I realize I'm a long way from the top—in fact, there is no summit! In science there are always new problems; if it weren't AIDS, it would be something else. I'm a gambler out for the big killing. Like a roulette player at the table, I'm addicted to getting results out of my laboratory. Last week everyone was away at a conference. With no one in Paris doing experiments, I got very nervous, like a junkie suffering withdrawal symptoms.

You've said many times, "I have lots of enemies."

I do! In France we're very egalitarian, so if you get out ahead of the pack, they shoot at you. I'm a target. This comes not only from my scientific success but also from my success in the media, which is something new for a scientist in France. From the start, AIDS has been a show-business disease. The press and media have been fascinated by it. People are making major discoveries in other domains, but they receive none of the attention accorded to AIDS, while I'm being barraged with invitations to appear on TV around the world. I've also made enemies because of the debate about who discovered the AIDS virus. You know about my affair with Robert Gallo, but there is also an internal debate here in France. All scientific research today is a collective enterprise, and some of my colleagues feel they haven't reaped the glory they merit.

To set the record straight, did you discover the AIDS virus?

There's no debate about this point. The argument with Robert Gallo had to do with proving causality. Did the virus I discover cause the disease? I don't think Gallo disputes that we were the first to isolate the virus and publish our findings in May 1983. All he has ever claimed is that he isolated the virus at roughly the same time. He wasn't able, however, to characterize it.

What was your reaction when Gallo announced that* he *had discovered the virus?

I remember quite well the day he came to my office in April 1984. He sat at this table, in the chair you're sitting in now, and told us he had discovered the virus that causes AIDS, which he was calling HTLV-3. It was obvious his virus was close, if not identical, to ours. My reaction was altogether positive. He was confirming our work. Afterward the debate became polemical, but my first reaction was, "Good, I'm pleased Gallo has rediscovered what we've already found."

Even though he was claiming all the credit for himself?

We both contributed to the discovery of the virus. The difference between science and religion is that in science everyone has to agree. For a fact to be a fact, it has to be reproducible. Miracles, by definition, are not reproducible. So if we were capable of isolating the virus that causes AIDS, it's not surprising that others could do it as well. For a long time Gallo rejected the idea that this was a new virus completely different from the leukemia-producing virus he'd discovered. It took him awhile, but he was finally forced to admit that we had something new.

What was Gallo's contribution?

He found a way to grow the virus in continuous cell cultures. We developed a similar technique at the same time, but our cell lines were less productive than his. Later we found one equally as good, but in the beginning his line was better. This was important for developing the AIDS blood test. We also owe to him the idea that AIDS was caused by a retrovirus. You have to distinguish between isolating the virus and demonstrating its causal role. In my opinion, it's more important to demonstrate causality. That's why I think Gallo and I made equal contributions.

Some people say that Gallo owes his discovery to samples of virus you sent him in July and September of 1983.

I don't want to stir up the past. All the details are given in the chronology we published together in *Nature* magazine. It says I sent him the virus. These shipments must have been useful to Gallo, and I don't think he denies it.

Is it possible that Gallo's cell lines might have become contaminated with your virus, which would explain why he reproduced it so faithfully? [In 1991 it was discovered that* both *Gallo and Montagnier's

laboratories had been contaminated accidentally by a renegade French virus.]

These accusations were made by the Institut Pasteur. And Gallo himself did not exclude this possibility, although he could argue that he also had an independent isolate from a Haitian patient, different from ours. But let me give you another example. When he was trying to isolate the second AIDS virus, HIV-2, Max Essex at Harvard apparently contaminated his cultures. What he called HTLV-4, using Gallo's terminology, was actually a monkey virus. This happens quite often in labs where scientists cultivate their cells in CO_2 incubators. This technique makes it impossible to keep infected cell lines completely isolated. While the gas is entering the incubator, microaerosols of virus are escaping. After some bad experiences, I rejected this technique for isolating viruses. Instead, we use a system of gas-filled bottles that are completely closed to the atmosphere.

Because of his ability to mass produce the virus, Robert Gallo has been called the Henry Ford of AIDS research.

Gallo is not someone who has merely perfected other people's discoveries. Many important findings have come from his laboratory, things like interleukin-2, the growth factor that allowed us to isolate the AIDS virus. He generates a lot of creativity. He's not merely a Henry Ford, a biological mechanic.

Gallo and I have worked together in the past, and we'll probably do so again. The unhappy period that he and I lived through was distorted way out of proportion by the press and by the politics of the disease. There was terrible pressure in the United States that an American be the first to discover the virus, while France was relatively disinterested. For a year we worked completely on our own, with no one understanding the importance of our findings. Gallo didn't believe me at the time, which put him in the enemy camp, but I have many rivals closer to home. In the end, Gallo and I have the same enemies, which makes us allies.

What was your reaction to the political pressures surrounding AIDS research in the United States?

I was particularly furious that our patent for the blood test was ignored until Gallo's was accepted. That's what pushed me into starting legal proceedings. Scientists in the United States are forced to produce results, which sometimes warps their sense of ethics. Scientists have even faked their experiments to look like winners, and not only in the United States. The best way to avoid this is to have several currents of thought and different countries working on the same problems. Maybe ninety percent of the world's biological research comes from America, but to avoid people's approaches becoming too rigid, it's important to have laboratories existing beyond the pressure of the Americans. This is why the AIDS virus was discovered as quickly as it was.

Were you surprised by the nature of American science?

No, I really don't object to the aggressivity of the Americans. I object to the passivity of the French, who met my work with incomprehension and indifference. Thanks to this research, France could be making breakthroughs in biotechnology, but it's letting the opportunity slip through its fingers. There is a disequilibrium between our scientific abilities and their industrial applications. This, not the pushiness of the Americans, is what frustrates me.

Like a lot of Americans, I'm a self-made man. There was no portrait gallery in my house. I'm the first person in my family to go to college, but I was born with this great will to succeed, to overcome what some people would have seen as handicaps. The war deeply affected me. We ate mainly potatoes, and I didn't gain weight for four years. I suffered from malnutrition, which weakened my health and made me sick a lot. My mother wouldn't let me go swimming because she was afraid I'd drown. I'm like those handicapped people who go on in later life to break world records.

Have you ever thought of moving to the United States?

I'm not opposed to the idea, but even if I did, I'd remain very French in my sense of measure, logic, and love of good food. I was born in the Loire valley, where people live reasonable, ordered lives. My school was called the Lycée René Descartes, and Descartes himself was born twenty miles away. Because of this Cartesian influence, the French are endowed with good sense. Alas, the United States is not very Cartesian.

People think the Institut Pasteur is rich and that I, its incarnation, must also be endowed with all the funds and equipment I want. This is not the case, and if the situation ever became impossible in France, I wouldn't hesitate to move. I want to keep working! I know what has to be done to conquer AIDS. I'm not doing this for personal gain. If it were only money I was after, I could exploit my renown and live off the fame of my past research. Quite the contrary, I live a hard life, with no vacations, short nights, and long days that are filled with thousands of things I don't have to do but I feel I should do. If I can do my work in France, I'll stay here. Otherwise I'll go elsewhere. I'm too young to be embalmed in a glass coffin.

Were you pleased with the legal agreement you and Gallo signed in 1987?

Yes. I thought from the start there had to be a compromise. No one should be made to look as if he were losing face. The only solution was to split the royalty money fifty-fifty and establish a foundation for spending it. I was probably happier about the settlement than Gallo, because it was my idea. Many people thought I could have done better, but I don't think so. The affair caused a lot of ill will, and AIDS is too important for the problem to have remained unsolved. It was giving certain scientists—and science itself—a bad name. Not to have fought would have created a bad precedent. It would have signalled that one can get away with anything in science, which isn't true.

Are you under a gag order that prevents you from talking about the details of the accord?

It's not exactly a gag order, although it's stated in the agreement that no one will reopen the scientific argument. There were actually two agreements: a legal accord between the American government and the Institut Pasteur, and a scientific accord between Gallo and me, which was published in *Nature*. The scientific agreement took a lot of work, and we only finished the task a few days after President Reagan and Prime Minister Jacques Chirac announced the legal settlement. I flew to Frankfurt and met Gallo at the Intercontinental Hotel on his fiftieth birthday. I took him a bottle of cognac, but Gallo said he wouldn't drink it until we were finished. We worked right up to the last minute before I left to catch my plane. So I never did get a drink of that cognac.

Now Gallo and I are getting along quite well. We respect each other. This often happens to people who've fought a lot. They finish with a better understanding of each other. Gallo and I were friends to begin with, and we've ended by being friends again. I bear no grudge against him. My rancor is reserved for the people who are still trying to get in the way of my research. I have a reputation for being an imperialist, an expansionist, because I ask for a lot of money. But this is what it takes to do research on AIDS. AIDS is not an affair that's going to last fifty years. It's going to be settled in ten years, and if you want to put the package together, you can't drag your feet.

Do you deserve a Nobel Prize for discovering the AIDS virus?

It's not for me to say. The Nobel committee might want to give the prize to the discoverer of the vaccine, although it was the discovery of the virus itself that allowed for its detection in blood and the development of public health measures that can limit the epidemic, even without a vaccine. The contribution of the American team is also important, so I doubt the prize will go to only one of the virus's co-discoverers. If someone develops a miracle drug against AIDS, that, too, would merit a Nobel Prize. It has already been five years since the virus was discovered. AIDS is a terrible malady, and I don't want to suggest that scientists are reaping their honors at other people's expense. I haven't changed because of my notoriety; but there's tremendous pressure from the media and the public, who think of us as a cross between magicians and movie stars.

Tell me about the research program you've recently launched.

This year, for the first time, we're getting substantial money from the European community. Research on AIDS involves more than conquering the disease. Many industrial spin-offs will fall to the pharmaceutical companies and biotechnology firms. Cetus, a San Francisco firm, has developed a machine that allows us to multiply a cell's DNA thousands of times over. The technique, called gene amplification, is crucial for a virus like AIDS, which can lie hidden—unexpressed and undetected—in the lymphocytes and macrophages [immune cells that engulf invaders]. The only way to reveal its presence is to find the DNA of the virus. But gene amplification is not only important for AIDS; it will be useful for detecting all sorts of genetic maladies. Cetus could sell thousands of machines and make a fortune from its invention.

Where does the AIDS virus come from?

Human beings had their origins in Africa, so it's natural that a virus associated with them should also have originated in Africa. The prototype virus has probably been in humans for a long time. We know this from looking at its evolution in different species of monkeys. There is a virus of green monkeys; a virus of mangabeys; a virus of baboons and mandrills. But all these viruses have the same basic properties as the human virus. Like keys fitting into locks, they recognize the same sequences on the immune system's T4 white blood cells. The AIDS virus may be as old as the evolution of primates, because the viruses diverged with the different species themselves. Although Africa is the likely source of AIDS, one could debate the point for hours. First, you have to disassociate the virus from the epidemic. The epidemic is new, the virus is old. But is the virus old in humans, or did it develop after a passage from monkeys to humans?

One could imagine a scenario in which the virus lay hidden for generations behind other diseases that killed people at a relatively young age. If so, what explains its sudden emergence in Africa and America? If AIDS is of African origin, why didn't it come to Europe before the United States? Historically, we've been much more closely linked to Africa. There was the American slave trade, but apparently the slaves—at least those surviving the crossing—didn't have AIDS. The virus seems to have come to Europe from the United States. It might even have reached Africa via the same route.

There are other hypotheses concerning the origin of AIDS, such as the "American" hypothesis. The traffic in blood might have caused the epidemic. An isolated population in South America could have been the focus of a natural infection that was amplified by the sale of blood products to the United States. These are poor countries, and a lot of South American Indians supplement their income by selling blood.

Why do you think the virus is old?

We're boarding a train that's already in motion. New species aren't being created. We're seeing the old ones evolve. The AIDS virus's complexity shows that it has undergone an arduous process of selection. With nine genes, it's the most complex retrovirus known. We find the virus now in many species of monkeys—a whole family of viruses. But it seems to have reached a state of biological equilibrium that keeps it from being pathogenic in mammals in the wild. It's hard to know for sure, since a monkey can die with no one seeing it, but the virus appears not to be lethal for green monkeys and mangabeys.

If you admit the virus is old and the epidemic is new, then it's not the virus that has changed its basic configuration over the years but the behavior of its host. Our civilization has created the epidemic. Of this I'm absolutely convinced. We're a civilization of blood, of blood transfusions. This practice has existed for only a little over half a century, and then came the sexual revolution. We've created one world environment for our germs. The globalization of culture has globalized our parasites.

You could say that AIDS is a disease of the Boeing 747. The big jets are its vector, and without them there would be no AIDS epidemic.

Do you think that monkeys passed the virus to humans?

This seems to be the case with the second AIDS virus, HIV-2, whose epicenter lies in West Africa. But it's not necessarily the case with HIV-1, the virus now spreading through the industrial world. That human AIDS comes from two different viruses is abnormal. HIV-2 so closely resembles one monkey virus that it could have been passed accidentally from monkeys to humans. But no one has found a monkey virus resembling HIV-1, with the exception of one study that seems to have found it in two chimpanzees. Green monkeys are innocent of giving AIDS to humans. This might be true of all monkeys, at least for HIV-1. Green monkeys are also innocent of giving us HIV-2, although other species, such as the macaque and the mangabey, might be implicated.

So where does HIV-1 come from?

Humans. It's a classical notion in virology that a change in species makes a virus pathogenic. This is also true of viruses that move into a different population of the same species. Perhaps an ancient strain of the AIDS virus, tolerated by an isolated African or American population, later escaped from this state of biological equilibrium to infect the world at large. A study supports this theory. In the villages of eastern Zaire, where the disease is pandemic, the percentage of people seropositive, or testing positive for AIDS antibodies, hasn't changed in ten years.

Why in the United States did the virus first attack homosexual men?

The AIDS virus plays the role of a lion hunting a troop of gazelles. It will bring down only the weakest among them. Likewise the virus will kill children and adults with immune systems less strong than the others'. The immune system of male homosexuals is already depressed. The virus searches for favorable terrain in which to establish itself. It creates an epidemic in territory already prepared by the cofactors that male homosexuality generates. Not only the establishment of the virus, but also its transmission is aided by cofactors.

AIDS is a disease of the cities. In the developed world, in Africa, it's the same thing. City living has created the kind of promiscuity that allows the virus to spread. Other cofactors are generated by city living. Environmental pollution can also depress the immune system. The fact that someone becomes seropositive for AIDS might itself be associated with cofactors. It's not easy to become seropositive. Recent discoveries have shown that the virus can exist in a latent state, unexpressed until it eventually breaks through our immune defenses. One can be infected without showing any signs of antibodies, which means more people have the AIDS virus, in its latent form, than the official statistics indicate. Clearly we're missing a lot of pieces in the puzzle.

Have you been to Africa?

I've visited the Pasteur Institute in the Central African Republic, one of the poorest countries in Africa, and Tanzania. My limited experience in Africa has already taught me a lot. Everything needed for transmitting AIDS is found in an African hospital, where conditions are unimaginable. If AIDS is a disease of the cities, it's partly because cities have hospitals.

Why is AIDS transmitted heterosexually in Africa?

Other cofactors are at work, including frequent genital ulceration and infections of women. Also female circumcision—clitoridectomy—favors infection by the virus and its transmission.

What are the differences between the two AIDS viruses?

HIV-2 is found in the old Portuguese colonies of Africa. The virus seems to have originated in Guinea Bissau and spread from there to the Cape Verde Islands and the Portuguese colonies of southern Africa. We first isolated HIV-2 from African patients dying of AIDS in a Lisbon hospital. So it was clear from the start that HIV-2 caused a disease as fatal as HIV-1. The two viruses provoke the same neurological disorders. But there's tremendous variability within each group of viruses, differences both in genetic variability and pathogenicity. Certain strains of each are more virulent than others.

How do you define a virus?

It's a parasite that can't exist without a cell. You might compare it to the cassette in your tape recorder. Without a machine to play it back, the virus is useless. Viruses are biological objects, but they're not living objects. All the genes in a cell are integrated into the proper functioning of that cell. But there is always the danger that a gene could escape from this integration and replicate itself rather than the DNA of the cell. This fragment wants to protect itself from dying; so it surrounds itself with a few supplementary genes and protective proteins. These allow it to be inserted back into the machine that's going to read it. This is a virus.

What distinguishes the AIDS retrovirus from other viruses?

Retroviruses are probably ancient genes that have broken away from the cell. They're primitive molecules trying to return home. Home in this case is the DNA of the chromosomes. While other viruses have developed a more independent existence, retroviruses have preserved the memory of their origins. A lot of retroviruses cause cancer in animals, and I suspected they might also cause cancer in humans. I thought I'd discovered one, a retrovirus that produces breast cancer, when my research was interrupted to begin working on AIDS.

Why do you describe the AIDS virus as intelligent?

It has a tremendous capacity for genetic variation. It plays roulette all the time, and it only keeps the good numbers. All retroviruses are highly variable because their enzymes have the intrinsic capacity to make lots of mistakes. But what's surprising about the AIDS virus is its ability to exploit this capacity. It leads a double life. Part of the time it has a vegetative, larval existence, like a cat that sleeps all the time—and when a virus sleeps, it doesn't mutate. But it also has a nocturnal life, when it wakes up and starts changing itself to resist the immune system.

The AIDS virus might have been vegetating for thousands and thousands of years until it found the civilization that spurred it into action. It's not impossible that social conditions analogous to our own provoked AIDS epidemics in the past. Promiscuous civilizations, with a lot of sexual contacts and changing of partners, could have ended in epidemics that killed a good part of the population. This could explain why all the world's great religions prohibit adultery. If I were a devil creating a malicious virus that would cause the most problems for the human race, the virus I'd create would be AIDS. Knowing people's capacity for making vaccines, this is the virus that has found the Achilles' heel of our immune system.

Gerald Meyers, at the Los Alamos National Laboratory, estimates the AIDS virus is forty years old.

He's dealing with current rates of mutation, when we already have an epidemic. But without an epidemic, the acquisition of changes in the genome can take place much more slowly. Meyers took the precaution of saying the virus is at *least* forty years old. Other people say that HIV-2 was derived from HIV-1 forty years ago, or vice versa. But HIV-2 has genetic and biochemical properties that make it very different from

HIV-1. It's closer to the simian immunodeficiency virus of macaques. This is why the divergence between the two is much older than forty years.

Can the AIDS virus be transmitted vertically in the genes that parents pass on to their children?

So far there's no evidence that the disease can be transmitted genetically among humans. But Michael Martin and colleagues at the National Institutes of Health have done a remarkable experiment showing that AIDS can be passed transgenetically among mice. After genes of the virus are introduced into the ovum, all the cells derived from this egg are infected. The baby mice have the AIDS virus throughout their bodies, and they die within thirty days. The virus normally seeks out two targets, macrophages and lymphocytes; the mice die of an infection in the macrophages.

Why is it that the macrophages have become a hot topic in AIDS research only recently?

It's partly the fault of my laboratory, I'm afraid. AIDS is essentially a disease of the lymphocytes. So that's naturally where we began looking for it. When you separate lymphocytes from macrophages, it's easy to lose the latter. We knew how to culture lymphocytes, but only two or three years ago did we learn how to culture the virus in macrophages. Gallo's lab deserves credit for this discovery. That the virus develops in the macrophages is crucial to explaining the neurological symptoms of AIDS [macrophages operate in the brain]. Infected macrophages secrete substances that poison the immune system. Lymphocytes die immediately after they're infected. But infected macrophages continue working as the reservoir of the virus.

The AIDS virus mutates so fast that doctors report cases of people dying from a different strain than the one that originally infected them.

Because the polymerase [an enzyme] of the AIDS virus makes ten thousand times more errors than a normal cell, it has ten thousand times the possible number of mutations. The risks involved in the disease can change in the middle of it. The tropism of the virus for the macrophages might depend on one specific mutation. You must also realize that someone with AIDS is infected not with one virus but a mélange of different viruses—a virus soup, with all of them helping one another out. This is a very dangerous situation because it can evolve in whatever direction it wants. For this reason I describe the virus as "intelligent": It "knows" the need for genetic diversity in assuring its survival. This is why racism is so idiotic, and a virus has finally reached the same conclusion.

Does this chameleonlike quality of the virus make it impossible to find a vaccine against AIDS?

We're changing our ideas about what's required for developing a vaccine. Initially we thought it could be made from the protein envelope of the virus, which varies the most. We also wanted to keep in mind the virus's highly selective attraction for T4 lymphocyte cells, a tropism that's probably relatively stable. But antibodies made against the envelope don't offer sufficient protection.

Now we're looking at proteins inside the virus to introduce what we call cellular immunity. We have fewer paths to follow, but one of them might be good. It's a gamble, but the discovery of a vaccine isn't essential to ending the epidemic. If the AIDS epidemic has its origins in the nature of our civilization, one can halt the epidemic by modifying this civilization through public health measures and education. Admittedly, this is a slow process that could take a lot longer than developing a vaccine. These changes can't be forced on us, as they were in the past, by social taboos and religion. There's no going backward. But people have begun to redefine the idea of love, and this could be quite fruitful.

How well is your research on the vaccine going?

It's not going well at all. We have to do better than we're doing, and we need an animal on which to test the vaccine before moving on to humans. We can't use chimpanzees. They're not enough of them, and they cost too much. So we have to find another monkey capable of being infected with AIDS, maybe the macaque.

What do you think of Daniel Zagury at Pierre and Marie Curie University, who has tested an AIDS vaccine on himself and fifty volunteers in Zaire?

I could make a joke about chimpanzees being rarer than humans. Zagury has had interesting results on himself, but he's well aware that he still doesn't have a vaccine. He wouldn't dare inject himself with the virus to prove he's immune. The bad news is that no level of immunity you induce in an animal is sufficient to protect it from the disease. The same is true of human beings. But maybe our test conditions are too severe. We inject the virus directly into the blood, but the natural route for transmission is sexual, where the doses are much weaker. Maybe we could find a vaccine that suffices for this kind of transmission, but it's going to be hard to prove it works.

Would you ever inoculate yourself with an experimental drug?

Without a doubt. I often give blood in the laboratory. Knowing that the risks were minimal, if I had to inject myself with something, I wouldn't hesitate to experiment on myself.

What do you think of Peter Duesberg at the University of California at Berkeley, who says HIV-1 is not the cause of AIDS?

Peter is playing the devil's advocate, which is an important thing to do, but nonetheless I think he's wrong. His objections were put to rest a long time ago. When I first isolated the AIDS virus, I wasn't able to say for sure it caused AIDS. There were a lot of reasons to think that AIDS *wasn't* caused by a virus; so I had to eliminate these objections one by one. I agree with Peter's argument that not everyone infected with the virus will get AIDS. But while Peter says there is *no* principle factor causing AIDS, I say the *virus* is the principal factor, along with a lot of cofactors. You can have all the cofactors, but without the virus, you won't have AIDS.

Do you treat AIDS patients?

I sometimes visit patients in Pasteur Hospital if we're running tests on them. People in the final stages of the disease resemble terminal cancer patients. Our progress in treating AIDS-opportunistic infections has led to our seeing a lot more people dying of the AIDS virus itself. It's agonizing to see someone wither away to a skeleton, and it pushes me to work harder on experimental treatments. I have no choice but to try everything possible. People have put their confidence in me. It might be misplaced, but they're waiting for me to do something.

The Centers for Disease Control [CDC] predict that AIDS has a fatality rate of nearly one hundred percent.

AIDS does not inevitably lead to death, especially if you suppress the cofactors that support the disease. It's very important to tell this to people who are infected. Psychological factors are critical in supporting the immune system. If you suppress this psychological support by telling someone he or she is condemned to die, your words alone will be condemnatory. It simply isn't true that the virus is one hundred percent fatal. If you lead a normal life—sleep regularly at night, avoid alcohol, coffee, and tobacco—your immune system could perhaps resist the disease for ten or fifteen years. By then we might have found an effective therapy. Furthermore, the CDC statistics are biased. They're based on a male homosexual population in San Francisco with a lot of cofactors aiding the disease. The same thing may not be true of young men and women now being infected. AIDS is not only a disease of the big city. It's also a disease that strikes a certain sector of the population—journalists, television personalities, artists, singers, actors, people who lead a certain kind of life. And this is no accident.

Were you ever warned against working on AIDS for fear of jeopardizing the reputation of the Institut Pasteur?

This didn't stop me for a second. I began working on AIDS because an affiliate of the Institut Pasteur was manufacturing an antihepatitis B vaccine from human plasma, and the scientific director wanted to know if his blood supply could be contaminated. Guaranteeing the safety of our vaccines got me interested in the problem. When it became known what I was working on, people at the Institute began to talk: "What's Montagnier doing looking at a disease of homosexuals and other marginal people? This is bad for fund raising." I was discouraged and demoralized by this reaction, but it didn't stop my going ahead, because I found the research itself exciting. I'm not a homosexual, and it was irrelevant to me whether I was researching a disease of homosexuals, drug addicts, Hindus, or whatever. Many scientists have an irrational fear of AIDS. The Institut Pasteur recently built some new laboratories, and they didn't give me one. I suspect this is because the other scientists were afraid of having the virus inside their building.

What are the differences between the United States and France in AIDS?

AIDS is an enormous problem in the United States, where it's the number one public health issue. This is not yet true in France, even if it is the European country with the most cases. The French are a Latin people who take their sexuality lightly. No one dies of sex; it's just not possible. And everyone assumes that if the Institut Pasteur is working on the problem, it will be solved in short order.

Do you segregate school children who have AIDS?

In this regard the French haven't panicked like the Americans. No one fears having the children of hemophiliacs in school. The extreme right has tried to make an issue of AIDS, but people haven't followed their lead. I was just traveling in the Middle East, where a lot of countries demand a medical certificate "proving" you're seronegative. But this strategy doesn't make any sense, especially in countries that are already deeply affected, like France and the United States.

Are people like Rock Hudson still coming to France for treatment?

Those days are over, and the drugs we were offering turned out to be disappointing. We're working with new drugs, but nothing that justifies a sick American coming to France. As soon as we find anything interesting, it will go straight to our American colleagues. No one has an interest in monopolizing a cure for AIDS. Our principal goal at the moment isn't to treat advanced AIDS cases but to find other medications, possibly very mild ones, for treating people who are newly infected. We've set up a screening unit for testing antiviral products. Automated and very fast, it allows us to test two hundred chemicals a week. And one of these days we might find something to get excited about.

How does AIDS compare to the other great epidemics in history?

People who died of the plague thought they were being punished by God. Today we know that AIDS is a virus, and we know how to avoid it. How many people will die before the epidemic runs its course? If three-fourths of those who test seropositive for AIDS get sick and die of it, this will be ten or twenty million people. The consequences will be most serious in Africa, where the disease will affect the economy and structure of society itself.

Albert Camus said that plagues and wars always take people by surprise. Were you surprised by AIDS?

I would have to say yes. But epidemiologists have known for a long time that we're vulnerable to new epidemics. The same civilization that created the AIDS epidemic could create others, with infectious agents even more virulent than the ones we know today. We haven't exhausted all the germs in our tissue capable of being transmitted by sexual relations. The greatest danger lies in nonconventional viruses that produce no immune reaction. They resist sterilization and all known drugs. Our civilization is in the process of selecting the successful germs of the future—those capable of escaping detection by the immune system. We already know that some brain diseases are cause by such agents. If any time remains to me after AIDS, this is what I hope to work on next.

James Black

✦ ✦ ✦

JAMES WHYTE BLACK has "an incurable mental habit of inverting every proposition" he meets. This knack for looking at the flip side of reality led to his inventing the world's first billion-dollar drugs. He did it twice: once when he discovered beta blockers, which prevent heart attacks and reduce high blood pressure, and again when he discovered cimetidine, the active ingredient of Tagamet, which cures ulcers.

Black is widely acknowledged to be the founder of the modern scientific approach to pharmacology. "He has designed drugs," says a colleague, "that behave like a rifle bullet, instead of a shotgun." On the verge of producing his third major discovery, the 1988 Nobel laureate in medicine says of himself: "I am addicted to medicinal chemistry. I get high on it."

Black is a deeply private man who resists speculating in public about the motives behind his research. We know he did medicine at St. Andrews in Scotland and then shipped out to Singapore to teach physiology in an old colonial medical school. He then spent eight years at a Scottish veterinary college developing the ideas that would revolutionize drug therapies for heart failure, high blood pressure, anxiety, ulcers, and other hormonal imbalances. We also know his father died of a heart attack following a harrowing car accident, and that Black's early experiments centered on angina pectoris, the crushing chest pain that can be precipitated by exercise and strong emotion.

Black saw his way to curing angina through a brilliant act of inversion. Our normal heart rhythm is maintained by a small group of pacemaker cells. These are controlled by neural hormones that release adrenaline into the blood and at nerve endings in the heart itself. Too much adrenaline can wreck the coordinated beat of the heart muscle and turn it into a twitching mass of unsynchronized contractions. Lack of oxygen reaching the heart is a primary cause of these contractions. But while his colleagues were looking for ways to force more oxygen through arteries narrowed by heart disease, Black took the opposite approach: Why not *decrease* the heart's need for oxygen by blocking the arrival of adrenaline?

Two types of adrenaline receptors, alpha and beta, were known to exist in blood vessels. But Black was the first to show that beta receptors respond to hormone signals in the heart. Next he set about finding a way to block them. Joining Imperial Chemical Industries (ICI) in 1958, he and chemist John Stephenson spent several years synthesizing chemical analogues to adrenaline. They were searching for an emasculated form of adrenaline—an adrenaline antagonist—that would bind to the receptor and thereby prevent the uptake of adrenaline.

Black knew he had found the drug he was looking for when he witnessed the first human trials with beta blockers in the early Sixties. A house physician at St. George's Hospital in London was injected intravenously with isoprenaline, a beta receptor stimulant. The man's heart rate and blood pressure shot up. His face contorted. His breathing became distressed, and he said afterwards that he felt as if he were going to die. But when the man was injected with a beta blocker, which was later perfected and released as the drug propranolol, the symptoms of his adrenaline rush were immediately suppressed.

Today beta blockers are used to treat a broad range of cardiovascular, kidney, and psychological problems. Because they alleviate the stress induced by too much adrenaline, they are favored by concert pianists, billiard players, and public speakers, Black among them. As he joked after hearing the news that he had won the Nobel Prize, "I wished I had some of my beta blockers handy."

Propranolol, sold under the trade name Inderal, was a gold mine for ICI, making the company billions of dollars. Black himself earned no great financial reward, and he soon quit ICI to work for a competitor, Smith, Kline and French Laboratories. He would make SK&F a billion dollars a *year,* by developing Tagamet, its best-selling ulcer drug.

Black discovered cimetidine by following the same line of reasoning he had used to develop beta blockers. After isolating the receptor that produces excess gastric histamine, he engineered a hormone antagonist that blocks its uptake in the stomach, thereby bringing relief to millions of peptic ulcer sufferers. But by 1973 Black was out looking for another job. He explains this pattern of precipitous departures as a defense mechanism against boredom. He wants to invent new drugs, not promote their sales.

Today he heads the department of analytical pharmacology at the

University of London's King's College School of Medicine and Dentistry and directs the James Black Foundation. Organized to design prototype drugs, the Foundation is being financed for ten years by Johnson & Johnson. Black's only regret about the billions he made for his former employers is how badly they spent it, with most of the money going into building corporate hierarchies. The best medium for new drug research, Black says, is small groups of twenty scientists. He has forbidden his Foundation to grow beyond this number and demanded that its royalty income go to support similar groups doing basic medical research.

I met Sir James—he was knighted in 1981—at his King's College office in south London. We later toured the guarded laboratories of the James Black Foundation. Casually dressed in a striped tie and red-checked shirt, Black was hiding the nerviness of a corporate executive behind the demeanor of a professor. Warned that my subject could be difficult, I was nonetheless surprised by the tirade that almost ended the interview after the first question. "Where were you born?" I asked. This was not an idle query, considering the variety of answers in print.

"Are you trying to analyze me?" demanded Black. "Am I on your couch to be dissected and poked? This sort of public stripping before we start, getting me on the couch and tearing off my protective gear—can you see why I'm allergic to it?" Black did agree to talk about his career in drug research and his scientific ideas in general. "While we're talking," he said, "I won't be able to suppress aspects of my personality. Things will come out on this tape I'll wish to God hadn't. You will eventually get things I can't suppress. Is that not enough?" It was.

✦ ✦ ✦

What clues do you use for discovering new drugs?

There's no way I can keep abreast of what's being produced. So I have to clutch to more general sorts of ideas. I use these ideas to drive myself. New drug development doesn't require omniscient people, but simply people trained to formulate the problem. We're not self-conscious enough. We're bewitched by the idea of discovery as a mental trick—so that if you could find out my mental trick, you could sell it. But it's not a question of mental tricks at all. It's a question of mental discipline, a systematic way the brain behaves that is subconscious in me. I can try to be self-conscious about it, but I have no idea if I'm getting it right.

Still, we can reduce the inefficiency in the process of drug discovery. That is one of my obsessions at the moment.

Like Friedrich Kekulé and the benzene ring, have you ever discovered new drugs in your dreams?

I don't think so, but I have absolutely no idea where my ideas come from. For all I know they come from comets. They creep up on me. The most creative thing your brain does is dream. It synthesizes totally new pictures of experiences you never had, manufactures them down to the smallest detail, and you are not in control of it at all. When I go to sleep, I try to determine what I'll dream about, because there's a continuity in the stream of thought. But I'll never find a program, like a piece of software, that will allow me to dream on command.

How do you propose to reduce inefficiency in the drug business?

I don't want to knock an industry that has been enormously good to me. Still, the industry doesn't do enough that's useful. There's a basic inefficiency in the process. Once the industry has something to develop—a molecule that stops nerve degeneration in diabetes or prevents high blood pressure, for example—it knows exactly how to get a product. It evaluates the toxicity of the compound, learns the most efficient way of making it, develops analytical techniques for following the fate of the molecule once it's given to someone. There's an enormous amount of work that goes into making the substance available for delivery to the patient. I assure you, you could make a really exciting film about the high tech behind the little pill. Unfortunately, the industry sometimes points the finger in the wrong direction when it talks about high research costs. It should really be talking about high development costs.

What's the relation between research and development?

R and D are totally different. D—development—is knowing where you want to get to and what you have to do to get there. But for this R bit—although you think you know where you'd like to get to, you're not at all sure how to get there. The development process responds to hierarchical management, to whipping, kicking, and pushing. But it's inconceivable that you can make research go that way. Still, you can't just walk away and leave it, because it would wander like water all over the floor. Research has to be constrained and channelled. But the more you try to make it efficient, the more inefficient it will become. You'll kill the goose.

Should drug companies separate research from development?

The R man, like me, would be crazy to run a big corporation. That's why my present passion is to split R off from D and make it a separate thing. R is an activity for small groups. They require management that appears not to be suffocating but nevertheless is quite focused and constrained. Because you're not sure how this is going to work, you can't put all your money on one group. So you want several small groups, which are easy to manage by this gentle technique and are suitable for investment, because it won't be a disaster if one unit fails. This gives you another advantage. You can see which group is better than the others. They're now competing for a place in the sun, and that's how research should be organized: "To him that hath."

After you developed beta blockers, didn't you have a violent argument with your bosses at ICI before leaving the company?

That is absolutely not true. There's a gossip machine out there generating all kinds of myths. I left because I lose interest in a project as it goes from R to D. It simply isn't appropriate to the way I manage myself. As propranolol was being promoted for the marketplace, it was natural that ICI would want my input—to go to symposia, talk to visitors. In no sense was this a cynical attempt to use me for publicity. It was simply natural, and I had seen it happen to others, like James Raventos after he discovered the anesthetic properties of halothane.

If I had made a big fuss and said, "This isn't what I want!"—had I been the kind of person I'm supposed to be—ICI would have tried to accommodate me. But it didn't happen that way. By accident, one of my chums was moving off to another company, and they were looking for a pharmacologist. I stunned him by saying I was interested. It made life easier for me to move off to a new start in a new project. It was an efficient way of getting a clean piece of paper. ICI, understandably, was disappointed that I left. But in no sense was there any violent disagreement or argy-bargy.

Had you stayed, would the company have financed your histamine receptor research?

How can I know? I tried to fly the idea. I didn't throw tantrums about it. But again, you have to understand, here's a company that's got what looks like a hot piece of property in its hands, and it wants to concentrate on it. This gets misread, as though they were purblind or wicked. I don't think it was like that at all. The managers I had, like Garnet Davie, were superlative people. He did for me what I'm now trying to do for young scientists. He gave me room for maneuver by keeping people off my back. The same was also true at Smith, Kline and French and the Wellcome Foundation. Let's get the record straight. Whatever company I have left, it was simply because I was motivated to do something else. Having said that, don't misunderstand me. When I'm in a company, I'm probably not all that easy to live with. I hustle for my point of view. I get passionate. But I think that's always been accepted, and my behavior has never been distractive or threatened the integrity of the company. At least that's how it seems to me. Maybe others have a different view.

You went to ICI with the idea of developing propranolol. How did you get the idea that such a drug was even possible?

I'm not at all curious about this, to tell you the truth, and I'm worried about the shallowness of my recollections. I have a set of memories, but I don't know whether I've just adapted them to be comfortable to live with. I find it more exciting to think about what I've learned than what I've remembered. In my Nobel lecture I try in a simple way to recall my learning about partial agonists and how slow I was to cotton on to the fact that this was my lifeline. I really wasn't very smart.

Where were you born?

Addingston, Scotland, in 1924. It's on the Lanarkshire coal field south of Glasgow. My father was associated with mines. What Addingston was like I can't remember. I've never been back. I was one of five boys. Now what conclusion do you draw from that? This is part of my private life, which I have to tell you is ordinary and boring. You can poke and pry as much as you like. There is no horrible big secret. We don't have lunatics in the family or anything like that. It's such an ordinary, boring old life.

Dullness can be revelatory in itself.

I don't particularly want to publicize the fact [laughter].

You were trained as a medical doctor and surgeon.

Yes, but in Scotland, the double degree doesn't mean a thing. It's what you get for doing the standard undergraduate medical training. My interest was, and I think still is, in being an applied scientist. For a number of reasons I got interested in heart failure. I saw it as a disorder of regulation, and from that I got down to the chemicals and chemical receptors. I realized that if I wanted to achieve a physiological change, I had to get myself a molecule with certain properties. I went to ICI and asked for support. They sent a very high-powered team, I realized afterward, up to the veterinary school in Glasgow, and as a result of that discussion, I was offered a job. They gave me a young chemist, John Stephenson, and that was the beginning of a wonderful period in my life.

Why were you working in a veterinary school?

I was interested in physiology. I got married. Assuming I lived on oxygen, it would have taken me about ten years to pay off my medical school debts. What does every young Scottish boy do when faced with this problem? He emigrates, for God's sake! So I went to Singapore, where I was able to keep faith with physiology by lecturing in a former colonial medical school. I was really just training myself, learning from my mistakes. So in that sense, I'm a primitive.

Then I was about to be a father and again desperate: no money, no job, no prospects. Just by accident I ran into my professor of physiology in Oxford Street in London, and he invited me up to the University of Glasgow veterinary school, where I was hired to start a physiology department from scratch. There was nothing there except an old building in Glasgow. So really, as a youngster, I got a free pitch to play on. If you're interested in physiology, you can't be uninterested in animals. The basic assumption in physiology is that no species is unique. It may be in its peripherals, but like motorcars, the combustion engine is still what it's all about.

Do you still do hands-on work in the laboratory?

I'd be quite unsuited for it. The work requires a high degree of manual skill in preparing the tissues we use. In my lab we may have sixteen different organ baths going at once, and each one contains a tiny little piece of tissue that might be two-millionths of a millimeter long by half a millionth of a millimeter wide. We have an experimental program designed to cancel out any variations between different pieces, different organ baths, different times of day, and so on. We have a complex pattern of dosing, with small amounts of chemicals put in at precise times. It requires a very disciplined regime. I couldn't do it. But I love to look over people's shoulders and get a vicarious thrill from it. This also influences my thinking.

Wasn't ICI chemist John Stephenson crucial in your developing the beta blocker propranolol?

Yes. He was a natural teacher who simply had to tell you about what he was doing. He had you all over his notebooks. This doesn't often happen across disciplines. I'd done some chemistry, but it didn't amount to much, and what John did was to make medicinal chemistry a living, breathing thing. I've now spent more hours talking to chemists than I care to remember, and enjoyed every one of them. Now that I have my own little team, I don't have to cross barriers to talk to them.

Chemists and pharmacologists are quite different kinds of people. Chemistry is a subject for discipline. If you do a reaction today, you can write it down and anybody can repeat it tomorrow. It's suitable for people with strong control needs. But biology isn't at all like this. You get astonished. You live with being surprised. You have to be comfortable with variation and ambiguity. Because we're such different kinds of people, it takes a long time for chemists and biologists to trust each other. This happens when you really feel there isn't anybody else who could do a better job for you than your mate.

Were you lucky enough to find a chemist as good as Stephenson when you moved to Smith, Kline and French?

John came to me by way of a Christmas present. He just arrived on the doorstep. But at SK&F it wasn't like that. I had to argue my angle. You have to remember I came into a company where people were happily doing their own thing. This chap arrives with a certain publicity, because of the beta blocking discovery, but who was I to tell them, "Let's do something else!" They weren't sitting there waiting for me to come and change their lives. So I had to argue I had a better way of doing things, but it was not immediately obvious or accepted. This is how I started working with Robin Ganellin. Once we got over the arguing period, we had a terrific time together.

You made some mistakes on the way to discovering the ulcer drug, didn't you?

Everybody makes mistakes. I've reported some of them, the big ones. Success actually teaches you nothing. It makes you think you're good, for a kickoff, and that's bad. Insofar as I've learned anything, my life is a long history of mistakes. Do you want a full catalogue of my mistakes?

Yes.

An experiment is a little window you create for yourself. Nature isn't out there fully displayed so you can gaze on it and take it all. In science we're trying to see beyond the visible, to see something that actually isn't there at all. You need your instruments, your technique, your chemical method. And you need a biological system to look at. These are the windows you peek through. Now if you've chosen the wrong window, you don't see anything at all. But you also have to be aware of how blind you are even when looking through the right window.

When you were developing cimetidine, why did you have to double back and redo your early experiments?

You have an expectation, a prejudice that drives what you're doing. For histamine at SK&F I was doing the same kind of operation that had worked for me with adrenaline. But I took the analogy too literally and rigidly. I hadn't yet learned that when you take a sound method and your expectations are not fulfilled, you have to retreat and say, "Wait a minute. What's wrong with my expectations?" I was still pushing, pushing, because I was so sure my image was right. It took me a long time before it gradually got through my thick head that we were attacking the wrong end of the histamine molecule. We were concentrating on the heterocyclic end, when we should have been attacking the amino side chain. That was the reverse of what we'd found elsewhere. I had to go back and find that a compound I had looked at four years earlier had everything I needed. I hadn't seen it because the window I'd provided for myself didn't allow me to see it.

Was this four years of wasted work?

That's not bad for a graduate course [laughter].

I understand it took SK&F fourteen years to develop cimetidine.

The stories about the development people at SK&F getting impatient with my research go back to the earlier days, when we were still going through our sterile period. The record is quite plain that I had people keeping other people off my back at SK&F, as I had at ICI. So yes, there were pressures to move into other areas and work on other people's ideas concerning gaseous secretions and antacids and so on. I started at SK&F in 1964, and the paper in *Nature* describing burimamide is dated 1972. Therefore, you know we had burimamide earlier than that. Already by

1971 we knew in principle it worked in humans, and by the time I left in 1973, we knew it was flawed, but we had other drugs coming on, cimetidine among them. We knew we had an effective drug against ulcers, but we didn't know just how big a bonanza it was going to be.

You have earned more money for drug companies—perhaps unwittingly—than any other individual on earth.

This whole D bit, getting the product out and selling it, has nothing to do with me. It's not my fault a lot of people have indigestion. I refuse to take the blame for that. For the same science, it could have been a much more limited disease. I didn't set out to make these companies vast fortunes. I simply wanted to solve the problem I saw in front of me.

What do I feel now? Neutral. I like to think that having done these things allows me to move on to other viable projects. At the moment I am anxious to find a way to fund basic biomedical science. This is a big challenge for us. The peer review system, both in England and the United States, is biased towards projects that have a large body of secure knowledge under them, where the person comes equipped with a proven track record. So it tends to favor orthodox science. There has to be a basic inefficiency about funding innovative scientific activity. The money has to come from people who understand there is no way you can know in advance which of these projects will be good.

Who is going to pay for this research?

Industry, particularly the pharmaceutical industry, is pumping a tremendous amount of money back into the universities. But this money goes where industry wants it to go. They want to buy things. They want jobs done. What they don't want is controversial stuff. I think we have to find a way where high-tech industries, which are often improving upon advances made in universities, funnel some of their profits, uncontaminated by utility, back into the universities. This little venture I've set up—the James Black Foundation—is an example. If by any chance we produce products that make a lot of money, then the company cannot grow, because I've put in an article that forbids it to employ more than twenty people. Any money over and above its needs goes to a trust, and that trust will be obliged to put its money back into basic biomedical research. I think we should explore all sorts of arrangements like this.

Were the billions of dollars you made for drug companies well spent?

There is an unspoken assumption that if we got a golden egg when we were smaller, think how much better we'll do when we are bigger.

Unfortunately, the linear regression is not positive; it's probably negative. The larger the company, the more difficult the process, at least in making these initial discoveries. So I would have been happier if they had put their money back into small research units and universities, but I'm sure all these chaps thought they were doing the right thing.

Why are therapeutic drugs so widely abused today?

People consume prescribed drugs because someone prescribes them. The medical practitioner is the narrow wicked gate through which all this stuff flows. Are people with this privilege and responsibility making wise decisions? It's made even harder by our susceptibility to the glossy vulgarity of advertising. I'd like to see more postgraduate training of physicians and different undergraduate medical courses in pharmacology. I want to see doctors trained for the drugs they don't yet have.

Why are patients so keen to take prescribed drugs?

We're living in times when many people are stressed. People are not leading idyllic lives. The more we adapt to living in cities, the more problems we'll have. Drugs are one way of helping people move from a less happy state to a more happy one. They're not a panacea. But if you're really ill, as I have been often enough, aren't you grateful for something that can relieve the pain or cure an infection or ulcer?

As for tranquilizers and sleeping tablets, both practitioners and the public are becoming less enthusiastic about their use. I meet a lot of people who take positive joy in saying, "I won't take drugs," as if there were something virtuous in not taking drugs. It isn't virtuous to be healthy. It's just a happy accident. So I think there are changes taking place. We require more convincing than ever before, both professionally and in public, about the marginal benefits of a new product. The industry and people like me are being challenged to justify in a quality-of-life argument what the benefits actually are.

Were you surprised that beta blockers turned out to have so many applications?

I remember discussing with ICI the claims they wanted to make in the literature describing propranolol. I did a lot of hustling to stop them from claiming the therapeutic benefits, like cardiac modulator, and have them claim instead its analytical property. This allowed physicians, as they learned about the biochemistry of receptor control, to think themselves into different situations. Because the drug wasn't described as a cardiac modulator, it was no big deal to try it out on something else.

They didn't have to change their programming, which comes from having claimed the class rather than the consequence.

What about the psychological uses of beta blockers?

One of the earliest subjects who got pronethalol, our original beta blocker, was a house physician at St. George's Hospital in London. He was the first person I saw being given intravenous isoprenaline, the beta receptor stimulant par excellence. Isoprenaline gives you many of the effects associated with what American physiologist Walter Cannon called the "fight, fright, flight" syndrome. The physician's heart was going fast and forcibly. He was bouncing on the bed and breathing as though in great distress. His face was contorted. Afterward he said he thought he was going to die, he had such a feeling of impending doom. Of course, when you inject a beta blocker, all is quiet and peaceful afterward. So from the beginning, I had this tremendous image of stress, and a way to relieve it. The stress probably had nothing to do with the isoprenaline getting into his brain. It was simply a distorted input of information. His heart was being driven as though he were running like mad away from some danger. But his muscles were being told they were having a little sleep. It may well be this disparity of information, when you drive one organ system out of kilter, that causes behavioral problems in human beings.

Can beta blockers be abused?

Snooker players, or billiard players, as you call them, take lots of beta blockers. But I'm jolly sure that no amount of beta blockers would turn me into a snooker player! Let's get that cleared up. If I were a professional player, how could I improve my game? I can practice like mad. But I can do other things and usually will. I'll smoke cigarettes or drink alcohol or coffee. I'll interfere with myself in a number of arbitrary ways to get myself psyched up.

Snooker is not exactly a normal psychological activity. My view is that if snooker players want to take beta blockers, let them, as long as it's known that anybody can take them who wishes. It probably makes very little difference to the outcome of the game, although it might help snooker players survive a few years longer. They must be suffering a great deal of stress when they try to pop that last black ball. Vast sums of money hang on it. The beta blocker protects that man against his stress. But it may also reduce his chances of putting the ball in. It might in fact make him a little sloppy in his play. But if he feels more comfortable and lives a few years longer, why begrudge it to him?

Do you take beta blockers?

I have done so, before giving lectures. The effect is very interesting. A number of concert pianists tell me they experience the same effect. They confront a more dramatic situation than giving a lecture, because they have to do it twice. You psych yourself up for the first half of a concert. It begins as a highly tense, passionate sort of thing. At intermission, you come off sweating, take off your shirt, and have a cup of tea. Then twenty minutes later you have to go back out and do it again. It's harder to psych yourself up a second time, but if you don't, your playing will be flat and insipid. Pianists tell me that when they take beta blockers they give a much more even and intellectual performance throughout. The same is true with lecturing. With a beta blocker, you lose all the heart pounding and dryness in your throat before you go on stage. You feel more comfortable, but there's a danger in becoming so relaxed that you don't put any passion into it.

You've said we're experiencing a drug explosion. What do you mean?

If you were a G.P. you'd know what the explosion was all about. There are tremendous things on offer. Biochemistry and molecular biology are making explicit the vastly complex system of hormonal interactions. Each day more and more hormonal substances are being recognized. We're even trying to exploit them by genetic manipulation. Temporary manipulation is bad enough, but permanent manipulation of our chemistry presupposes a degree of knowledge that may attain the dimensions of hubris. The industry has shown its technical competence. We'll find better ways of identifying the firing end of large molecules. Then we'll make smaller molecules that are easier to handle but do similar jobs—like propranolol. We're getting more and more cocky about what we can do. The problem will not be our ability to do things, but to carry the public along with us.

How does public opinion affect drug research?

We should only prescribe one of these poisons when we think the alternative is worse. The public at the moment feel aggrieved if they take a drug and are damaged by it. But if the majority are to enjoy the benefits of drugs, I fear a small number of people will be damaged. This is the price we pay. There is no way to avoid it. It is entirely analogous to driving a motorcar. For the majority of us to enjoy the benefits of easy transport, a few people are going to die everyday on the roads. But at the moment there is no rapport between the public and the drug industry. They are at war with each other.

What miracle drugs do you predict for the future?

As I said, we're getting cocky about our ability to understand hormone chemistry. Using high-speed computers and quantum mechanics, we can take all these atoms and bolt them together in different ways. We are moving into an age where synthetic chemistry is model driven. We're arguing that the problem set doesn't know any more about chemistry than we do. I doubt we'll be able to make some entirely new molecule unrelated to anything we've ever seen before, but I'd be the last to reject this knowledge were it available to me. Scientists have to be a bit cocky, because we do things that are really crazy. In pure science you don't even know where it is you want to get to. So what makes you think travelling there is going to be such a hell of an interesting thing? The fact is, a lot of scientific practice is tedious. And what makes you keep doing tedious things, other than this pain in your head?

Are there many more receptors to map and model?

Gosh, yes! But it's not just the odd receptors we should be mapping. It's the complexities in the way they're organized. This involves a hierarchical, multiplex system of control that we're just barely beginning to understand. I'm searching for mathematical ways to handle complex hormone interactions. Once we know how to handle these interactions, then the possibilities for discovering new, highly selective, drug actions are almost endless.

How do you use your knowledge of hormones to design new drugs?

There's a lot of information in the chemical, but there's probably as much information in the manner of its delivery, and I'm thinking about its delivery in time. This might involve a pulse or an amplitude change in concentration or the arrival of more than one chemical message simultaneously, and it's an extremely difficult task to produce this delivery. We're constantly trying to find better ways to mimic the provisioning of hormones.

I take this chemical and modify it to produce a related substance, which the cells still recognize, but against which they are no longer effective. I emasculate them. The important thing is to pick systems that illuminate a clinical problem, which means you know exactly what you have to do to get back into the clinic. Part of the problem today is that a lot of pharmaceutical research is devoted to identifying hormonelike substances and then producing the related compounds that manipulate the activity. But the research didn't start off disease-orientated. So you get the old joke about a drug looking for a disease.

Many diseases today, even when they appear to be caused by a bacterium or a virus, can be seen as breakdowns in physiological regulation, which might be managing our immune system, for example. It's my obsession that physiological regulations are all and only about cells talking to each other through chemical messages. These chemicals are analogous to drugs. You give someone a drug and hope it communicates a message at the chemical level. You're looking for a change in physiology, an ordered, integrated change that undisturbs the patient. This is what hormones do.

What is a hormone?

There is no agreement among us about how to define the word. It originally referred to blood-born chemicals secreted by cells. So some people would say that secretions released by a nerve ending were not hormones, but neurotransmitters. Associated with this enormously varied class of chemical substances—the hormones and neurotransmitters—is a class of proteins that are their conjugate receptors. Again, nobody agrees how the word "receptor" should be used. Is a receptor any site where a drug acts? Because I want to tie the words "hormone" and "receptor" together, I would argue that you should define a receptor by the hormone that activates it.

What have you learned about hormones and receptors from your experience designing new drugs?

There is an axiom in pharmacology that no drug can have a single site of action. You might produce a substance and claim it has a single site, and I will add, "Only as far as you know at this time." The point is to get drug classifications that are physiologically based, that follow rules and don't allow classification unless you have the evidence. The classification determines whether or not I have a legitimate right to expect that I can find a new drug. I was at a meeting yesterday when somebody said, "Wouldn't it be nice if we could find a drug that would enhance calcium recapture by muscle fiber?" But are we entitled to expect that this wish could be the basis for a highly selective drug action? My kind of pharmacology is interested in drug behavior at the chemical level. I describe drug interactions in thermodynamic terms. The work I do may have to be revised. It might turn out to be inaccurate. But the ideas I'm talking about ought to be timeless.

Are they?

A great deal of pharmacology operates now at the molecular level. How is it that channels open and close in the membrane? Our ideas about molecular interaction are impermanent, like our ideas about the structure of the atom. There will always be some smaller dimension of scale or time that technology allows us to explore. So if I say a drug is a hormone receptor antagonist, I'm really saying the drug *behaves* as though it were. This was the method I used in 1960 when I classified pronethalol as a beta receptor antagonist, which I don't think anybody would say was an inappropriate classification. But *how* this drug interacts with receptors, good heavens! Our ideas are changing all the time. Looking back, what I thought in 1960 was absurdly naive, and in ten years' time, if I'm still around, I'll look back on today and say, "My God! Did I really believe that?"

What are you working on now?

Two things. Here at King's College we're trying to deal with basic questions about hormone receptor systems. I'm trying to find mathematical ways to handle complex interactions between hormones. Over at the Foundation I've picked one hormone receptor system to study and have one or two others mapped out. The pattern is the same as I've used in the past. I've picked a critical problem I can illuminate by looking at it as a disorder of physiological regulation. So it meets all my criteria.

What's the critical problem?

I'm not going to tell you. Everybody who asks me I refuse to tell, not because there's anything to be secret about. But if I tell you and you write about it, because my research on two previous occasions has made a lot of money, the question then becomes, "Is this the next discovery that's going to make a lot of money?" I had no idea about those, and I have no idea about this one. All I know is that if I start talking about it, the problem will be given undue weight.

After winning the Nobel Prize, is it true you ran out of your office and went to the local pub?

Absolutely. I've had a history of winning prizes, and people aren't backward about telling you you've been nominated. But it was getting embarrassing to be introduced at lectures as the fellow who *should* have had the prize. Insofar as I did think about it, which wasn't very often, I just dismissed the idea, because I didn't seem to fall into the right categories. I do industrial applied science, which didn't look to me like the kind of thing that was Nobel Prize-worthy.

When you ask people, "How did you hear you'd won?" they always say, "We had a phone call in the morning from some journalist." So when my secretary, a temp, put her nose around the door and said, "Somebody from Reuters is calling for information about a Nobel Prize or something," I was pretty sure I was in trouble. "Tell them I don't know anything about Nobel Prizes," I said. Then I turned to my colleague Paul Gastovich and said, "Let's get the hell out of here." We sat in my car listening to the radio, and oh, God, there it was.

You see, someone in my position knows these prizes are double-edged. Many people organizing symposia and lectures need speakers. And so they look around for somebody who is prepared to relieve them of their anxiety. I'm in the business of relieving them of their worry, but I feel I'm just being used. I now have limitless opportunities to waste my time on small bits of personal publicity. Why should I want it? I'm not doing any more interviews. You're the last.

Thomas Adeoye Lambo

✦ ✦ ✦

SAVAGES are happy. They laugh and dance and forget their problems in the blink of an eye. Or so said the missionaries who first penetrated into the interior of Africa. It took an African psychiatrist—the continent's first—to explode this myth of the happy savage.

When Thomas Adeoye Lambo looked into the villages of his native Nigeria, he found plenty of psychotics and schizophrenics. In fact, the per capita incidence of mental illness in Africa is the same as New York City. But Africans treat crazy people as part of everyday life, and this high tolerance for aberrant behavior is what made the *un*happy savages invisible to Western eyes.

Lambo also discovered that African village life, with its strong tribal and familial bonds, has therapeutic benefits of its own. Employing what he calls "methodological syncretism"—the fusion of Western and traditional ideas—he began incorporating family members and villagers into his patients' psychiatric cures.

The missionaries had made another mistake in dismissing Africa's traditional healers as "witch doctors." Lambo found them employing many of the same psychiatric techniques he had learned at the University of London. Centuries before Freud, traditional healers invented the "talking cure," free association, group therapy, and behavioral modification. They also had an extensive pharmacopoeia of herbal and psychotropic drugs.

"Their psychotherapeutic sessions were vastly superior to ours," says Lambo. "They showed us we hadn't got it right." He borrowed liberally from these traditional healers while developing his village-based cure for mental illness. Faster, more effective, and one-fifth the price of a Western cure, Lambo's model has since been adopted by sixty countries throughout the Third World.

One of more than thirty children fathered by a Yoruba chief with twelve wives, Thomas Adeoye Lambo was born in Abeokuta, Nigeria, in

1923. His early missionary schooling included lighting bonfires on Sunday to burn up statues and masks seized from neighboring villages. He studied medicine at Birmingham University and then went on to get advanced degrees at London University's Institute of Psychiatry.

In the first of several famous research projects, Lambo was hired by the Nigerian government to study mental illness and nervous breakdowns among his fellow African students in England. Touring the wards, he discovered that sick Africans, in spite of their Ph.D.s and Saville Row suits, cast their delusions in terms of witchcraft and juju. Lambo already suspected that only an indigenous African psychiatry could deal with the African psyche.

Returning to Nigeria in 1950, he was appointed director of the Aro Hospital for Nervous Diseases, Africa's first mental hospital. Lambo's British wife, while waiting for the buildings to be finished, suggested he billet his patients in neighboring villages. This gave Lambo his original idea for deviating from psychiatric orthodoxy. Colonial administrators looked aghast at his next experiment. Using his own money, Lambo hired a dozen traditional healers to practice alongside his regular clinical staff. For twelve years he filmed and analyzed these "witch doctors" at work. "The major factor in their success was the incorporation of the family into their psychotherapeutic sessions, something that psychiatry in Vienna and London had totally failed to do."

Along with his experiments in traditional medicine and village-based psychotherapy, Lambo began studying the psychological effects of modernization in postcolonial Africa. Depression, anxiety, and other neuroses are the price paid for social change. In its "malignant" form, this anxiety spawns secret societies devoted to ritual murder.

In the late 1950s, Lambo invited colleagues from Cornell University to join him in the first comparative study of mental illness in the Third World. Working with control populations in New York City and Halifax, Nova Scotia, they showed that Africans and Westerners alike suffer from mental illness, although the symptoms are specific to the culture out of which they arise. Their book, *Psychiatric Disorder Among the Yoruba,* is now a classic text on African civilization and its discontents.

After founding the modern practice of psychiatry in Africa, Lambo ascended into the ranks of the continent's few internationally known scholars. He became head of the department of psychiatry, dean of the medical school, and vice chancellor of the University of Ibadan. In 1971 he left Africa to work at the World Health Organization in Geneva, emerging by 1975 as WHO's Deputy Director-General. Now retired and living in Nigeria, the grand old man of African science still travels the world advising everyone from popes to presidents.

What's your opinion of the term "witch doctor"?

It's a derogatory term coined by the missionaries. When I went to mission school, every Sunday we were sent into the villages and told to collect all the idols and carved objects that now fetch millions of dollars at Christie's auction house. We would pile them in the middle of the village and burn them. This was part of our mission to convert the savages to Christianity. But just as there is no one single religion, so too is there no one single way to practice medicine. I could call quite a number of modern physicians "witch doctors," such as the ones who do "exploratory" surgery so they can hand you a heavy bill.

Are some traditional healers better than others?

Of course. Some are generalists, while others specialize in treating blindness or mental illness. They may be powerful in one area, like herbalism, but they usually employ other methods as well, like the Ifa oracles. The Ifa spirit is a powerful thing. It's super-mathematical. You throw sixteen palm nuts from hand to hand, while marking the message of the deity in a sand-covered Ifa tray. These marks relate to one of two hundred fifty-six verses in the Odu, or Ifa oracular system. After twenty or thirty minutes of calculating, the Ifa spirit begins talking to the patient, telling him what's wrong with him.

How do the oracles work?

When I first began watching traditional healers talk to the Ifa spirit, I tried to be skeptical. Did they cheat? I don't think so. How accurate is it? I think it's fairly accurate. Probability and chance play a great part in medicine. We "real" doctors are often blamed for what we had to do and praised for what we didn't do. The fact of the matter is that your guess is as good as mine. But what's important is whether or not the patient believes you. When the patient and psychiatrist agree, you have positive transference. If the patient is skeptical about what you're saying, the transference is negative, and you might as well send him somewhere else.

Tell me about your work as a psychiatrist.

"When I first saw you, I knew I was going to get better!" Many patients have told me this. They were reacting to my stature, personality, charisma, influence, transference—whatever you want to call it. The same goes for anyone in a position of authority. You can be the world's greatest genius, but fail at your task for want of the right personality structure.

Are you a powerful healer?

I think I am a good healer. I recognize now that I have genuine love for people. I can meet someone for the first time and have this powerful transference. After working as a doctor in England for two years, I was one of eight doctors appointed to government service in Nigeria. The queue outside my consulting room was always the longest. It was not because I was brighter than the other doctors. It was because of my personality.

How did you develop your bedside manner?

Empathy, it's either in you or it isn't. I feel comfortable in any community, white, black, brown, or yellow. I have more honorary degrees from America than any other country, even my own. I was the keynote speaker at seven meetings of the American Psychiatric Association. Tom Lambo was their blue-eyed boy! This occurred at a time of tremendous tension in black-white relations in the United States. People were afraid to invite me into their homes, but I empathized with both sides in the dilemma.

What was your family like?

My father was paramount chief in the ancient town of Abeokuta. He had twelve wives and well over thirty children. But my "family" is even larger than this. In Africa, especially among the Yorubas, a child has no one single father. Your mother's brother is also your father. Psychodynamically, this substitution is a good thing. It allows for extended care and a choice of role models. My grandmother carried me on her back for many years and gave me her own breast to suckle. There was nothing in it, but it kept me quiet! I actually spent my early years thinking *she* was my real mother. We lived in a large compound, and even today there are people there whose relationships are so ill defined that I can't trace them. Only in the big city do we use the word "cousin," so everyone is known as your brother or sister.

My father was a farmer of kola nuts and cocoa. My mother was a cloth trader who ran a big shop employing quite a number of people. Nigerian women in those days had tremendous economic power, and even today there is hardly a market in West Africa where you won't find Yoruba women traders. My "brother" Joseph Lambo—he is actually my cousin, although he gets upset with me when I call him that—is a traditional healer. Even back in grade school he was interested in herbalism, which he practiced on all of us. But now that he's famous, he wouldn't do it for free!

Does he have magical powers?

Traditional healers tend to think that everything is supernatural. Africans on the whole still believe this. When confronting our destiny and unknown forces in the world, we are essentially a religious people. If my cousin is going to cure you with the leaves of a particular tree, he'll visit the tree early in the morning, chanting incantations and invoking the spirit of the tree. He also uses a great deal of psychotherapy. He looks into the coals of a fire to analyze your dreams and makes startling observations. "Did you ever do such and such?" "Why, yes! How did you know?" He sacrifices chickens and goats and uses their blood to wash a man's head. These sacrifices and ritual expiations to the spirits are very important to the cure. Psychotherapy takes many forms.

Why do you call this psychotherapy?

Because I am a psychiatrist, and psychiatrists don't do any more than this, even in New York City. In fact, they do less because they have no time for you. Traditional healers spend a lot of time talking to their patients, getting medical histories. They hold psychotherapeutic sessions, either jointly or in groups, where they analyze dreams, dance, or perform ritual sacrifices. They burn incense and commune with spirits. They go into the patient's home and place fetish objects in the corners to ward off evil spirits.

Unfortunately, the modern economy has made many traditional healers into money grubbers. But the ones I employed at Aro were serious people, and they were poor. At the end of his cure, a patient might give them some yams, a couple of chickens, or a goat. They weren't in it for the money. Their knowledge was passed from father to son in a seven-year apprenticeship. No one called himself a therapist just because he'd spent three years getting a B.Sc. degree.

How effective are they?

At Aro we taped everything they did for nine years, and we found their techniques to be remarkably effective. For both human and scientific reasons, I wanted to provide alternatives to Western medicine, which is not the only way to a cure. Patients coming to the hospital were asked whether they wanted to use Western or traditional methods. So the two developed along parallel lines, like the Chinese system.

When did traditional healers start working at the hospital?

They never worked at the hospital. In fact, the colonial government officially disassociated itself from my work. They sent me a letter saying, "We have just built the most modern psychiatric hospital in Africa. What are you doing hiring witch doctors? If one of your schizophrenics goes and kills someone, His Majesty's government will take no responsibility."

Aro was the first psychiatric hospital in West Africa?

It was the first psychiatric hospital in the *whole* of Africa. There were asylums for locking up the insane but no treatment centers whatsoever. All the nurses and occupational therapists I hired were Europeans, because it took years before Africans were trained to assume these positions. People still come to Aro from as far away as Tanzania and Botswana for a degree in psychiatry.

Why did you decide to lodge some patients in the neighboring villages?

I'm restless. I'm willing to take risks. I wanted to prove that psychotic patients aren't any more violent than normal humans. Their violence is caused by the way you look at them. You expect them to be violent, so even before they do anything, you tie them down. But for forty years now the Aro village program has worked. I used my own money to set up the experiment, and it proved my point quite conclusively.

Where did you get the idea of a village cure for mental illness?

From my wife, who is an historian and educator. We met at the University of London. She has tremendous empathy, and I am very lucky that I married someone who shares almost everything I stand for. When we arrived at Aro, the hospital hadn't been built yet, and the government told me to supervise the project. "You're not the administrative type who is going to be happy sitting down at a desk for three or four years," she said. "It will drive you crazy."

"You are quite right," I replied. In the evenings after work we would drive our Volkswagen into different villages I was curious to see. One night I got out of the car to speak to a group of people. When I got back in, my wife said, "They love you. They know about your work at the hospital. Why don't you ask them if you can billet patients here until Aro is built?" So that's how it started, although of course we kept using the villages even after the hospital was built.

When did you notice that village life itself had therapeutic benefits?

Long before the hospital started to function. The nurses were sent out into the villages at nine in the morning to make their rounds. By eleven they were back reporting to me on what was happening. Then I went out to have my "ward rounds" in the villages. "What happened yesterday, Mama?" I would ask someone. "My son is much better," she would say. "He went for a long walk and came back exhausted. He slept very well, and he's no longer talking in his sleep." We began documenting thousands of similar case histories.

How did you decide whether to put a patient in the hospital or the village?

I'd put one schizophrenic or psychotic in the hospital and one in the village, at random. I insisted that those going to the village be accompanied by their relatives. The others went into the hospital alone, just like in New York or Chicago. I wanted to prove that the village cure would be faster, better, with fewer relapses. Rehabilitation would be smoother, because the patients were accompanied by their parents. Even during psychotherapy, the mother or aunt would sit with us. This meant that when the patient was discharged, there was no special follow-up to explain to the parent what was being done.

What kind of psychotherapy did the village patients receive?

Some was planned, some spontaneous. This included talking to normal villagers. Communication is important. You don't want to isolate patients by locking them up. This leads to convulsions and heavy reliance on psychotropic drugs. These drugs were just coming onto the market in great variety and abundance when I started working at Aro. But I thought the most important thing was interaction with normal human beings.

How did the villagers treat their crazy guests?

I billeted patients in the homes of people who showed the greatest tolerance for these pre-literate, or illiterate, patients. Even today Africans are tremendously tolerant of what Westerners call "deviant" behavior. One reason why the myth of the "happy savage" developed is that no one noticed the *un*happiness in African villages, where psychotics and schizophrenics are treated as part of everyday life. This made it very difficult for us to diagnose the general level of mental illness in Africa.

Is this tolerance true of other agrarian societies in Asia or China?

Yes. Only Northern European cultures make clear-cut distinctions between normal and abnormal.

How did the village health care system work?

We paid five shillings a night for each patient and another five for each relative. In addition, the villages were given electricity and piped water to improve hygiene. The people living there were farmers, fishermen, and small traders. At first they were afraid I was going to put into their homes schizophrenics and depressives who would endanger their families. It took a year and a half of negotiations to set up the experiment, but finally we got everyone solidly behind the project, so much so that new patients could be taken into the villages at midnight and people would open their doors to them. During the twelve years I was there, not a single incident took place.

Which worked better, the villages or the hospital?

Unless they chose to be treated by a traditional healer, patients in both the villages and the hospital had planned therapy, injections of psychotropic drugs, electroshock, and so on. The only difference lay in the social dynamics. Those in the hospital couldn't talk to anyone but their psychotic neighbor, while those in the villages, after getting their shots of thorazine, could go sit in the market and talk to anyone they wanted. The village cure was qualitatively better. Even people who didn't recover totally were able to function on their own. They were less dependent on psychiatrists and nurses. They showed significantly lower rates of recidivism. There were also economic benefits. The village cure cost one-fifth of a hospital stay.

What other benefits come from a village health care system?

Psychotics don't regress to the bottom of the heap, as I saw in the big hospitals in London. People buried away in the depths of the hospital could no longer put on their socks and shoes. But you don't get this regression to the infantile stage in people who are maintaining social contact. Locking up patients for twenty or thirty years costs the government a hell of a lot of money. It's in everybody's interest to get these people out and productive as quickly as possible.

Are the villages still functioning?

At a low level. People have told me my experiment worked only because Nigeria at the time was in a preindustrial stage of development. Once a country is industrialized—people living in nuclear families in city apartments, moving here and there at the whim of their employers—it becomes difficult to tie yourself down caring for relatives. Eighty percent of Africa is still rural, but maybe as soon as the next generation it will be detribalized and industrialized. The hypothesis I tested so successfully at Aro may not survive these developments.

How many villages are still operating?

Four out of the original twelve. When I went to Switzerland in 1971, I said, "Psychiatry is like a religion. Either you have disciples who share your views, or you don't." All the boys I trained as psychiatrists prefer to work in white coats with stethoscopes around their necks. They probably think "the great master" is around the bend. The villages are maintained only because I left them endowed with money from the Ford and Rockefeller foundations, who put millions of dollars into this.

How have the villages changed over the years?

I began my experiment before the days of heroin, cocaine, and crack. It's a different matter now in Africa. We can't do anything with the young people on drugs. I wouldn't even be brave enough to put them in the villages. They'd ruin the entire social fabric. They're under such tremendous pressure they'd practically kill to support their habits. Nobody at present has the answer to this problem.

Are you still treating patients?

When I retired from the World Health Organization and went back to Africa in 1989 I said I wasn't going to touch anybody, except maybe some colleagues my age. But they bring me their sons and daughters, many of whom are in trouble with drugs, and what am I supposed to do? I send them to Aro, which has been designated a WHO training center for research and treatment in narcotics. They break windows, bribe nurses to sell them drugs. Now that Americans and Europeans are becoming health conscious about smoking and cholesterol, your cigarettes and butter are being dumped in the Third World. The same thing is beginning to happen with drugs.

How many countries have adopted your village mental health care system?

About sixty, although some of them have different orientations. They use halfway houses, for example, from which people go out to work. It's like the Christian church, where you have all sorts of different sects. The major thing is to incorporate the family, something that psychiatry had totally failed to do, because of the circumstances of its founding in industrialized Europe.

How did you find the traditional healers who worked at Aro?

Patients were being brought to me from different parts of Nigeria and as far away as Ghana, so I knew I needed a variety of healers. I spent six weeks driving around Nigeria, staying three or four days in different villages, watching the healers work. Later I wrote to the best ones, or sent messengers, offering them free housing near Aro and a monthly salary.

They used different forms of psychotherapy and dream analysis. They sacrificed animals and expiated spirits by killing goats and washing people's heads with various concoctions. To cure hallucinations, they used ritual scarification and put certain potions in the cuts. I would say that most of what they did didn't work. We weren't able to see any benefits. But their psychotherapeutic sessions were vastly superior to ours. They showed us we hadn't got it right.

Why not?

Because they had time for people. A traditional healer could spend hours with a patient and his or her relatives. For example, a young man was brought to us with his hands and legs tied up after a schizophrenic episode. The healer said, "Take off his ropes, and we'll watch him." They unbound him, and the boy didn't do anything violent or aggressive. Then he was given a potion made of ground-up leaves. No Western drugs were used on the patients being treated by traditional healers. They were in charge from beginning to end. The young man slept for two days. Later, when I had these leaves analyzed, I found he had been given a strong dose of tranquilizers and psychotropic chemicals. While the patient slept, the healer interviewed his parents. The boy stayed only nine days before he was completely recovered.

To Western psychiatrists, the diagnosis in this case looks simple. "The boy had a spontaneous remission." But I witnessed traditional healers handling hundreds of acute cases the same way. Their management was superb. Their patients were usually discharged within a month. If I had admitted that boy into the ward, he wouldn't have been released in nine days. His illness would have been aggravated to such an extent that he would have been there six months.

Are many psychiatric "cures" really spontaneous remission?

It's the secret weapon in our medical practice. We use it a lot. Not every illness has to be taken to the doctor. But in this particular case I don't think the charge is true. I worked with these traditional healers for nearly fourteen years, and I still visit them to ask how they're doing, how are their practices, and so on.

What did your Western doctors and traditional healers learn from each other?

I introduced into Africa this particular form of methodological syncretism. I arranged the marriage of traditional and Western cultures. The traditional healers themselves are now using tranquilizers, thorazine, and other psychotropic drugs, combined with psychotherapy, ritual killing, and the interpretation of dreams. They're even giving antibiotics in cases where patients have pneumonia or chest infections. I've been able to persuade them that only in extreme cases should they restrain their patients. They're becoming more modernized. They realize they can syncretize both approaches. Just as there is no one religion—there are many religions—so, too, is there no one medicine. There are many medicines.

What are the psychological benefits of ritual sacrifice?

I once treated a Cambridge-educated judge who ran his car off the road on his way to court. He was only slightly injured, but badly shaken when they brought him to me at the University of Ibadan. I examined him and gave him tranquilizers so he could rest. A few weeks later he came to see me. "How is your recovery going?" I asked.

"Tom, to be perfectly honest," he said, "I think this was a case of juju." By juju, he meant black magic. "I had a vision in which I saw my grandfather. He told me that in order to break the spell I should sacrifice a goat. Not believing in such things, I told my parents to sacrifice the goat, and you know what? Since then I've been as right as rain!"

This shows that in spite of our sophistication, our Cambridge degrees and Western ways, there remains down in the soul a belief in the metaphysical and the supernatural. This was why he was so agitated when I first saw him, but he couldn't tell me.

What other experiences have you had with ritual sacrifice?

I've seen hundreds of cases like this. The Nigerian government employed me to lead a team of researchers studying mental breakdowns among Nigerian students in England. There were twelve of us, anthropologists, sociologists, psychologists, and psychiatrists like myself. We went to every major hospital in the country. What interested me was the fact that people—despite their M.A.s and Ph.D.s—cast their delusions in terms of their African culture. They believed some sort of psychic ray or beam had come from Nigeria, from their mother's uncle or whomever, opposing their wish to become a lawyer or doctor. Some of them had been in England for years, but in spite of their long exposure to Western civilization, when they became ill they fell back on their culture.

Have you ever practiced ritual sacrifice?

Never. It's not that I don't believe in it, it's just that I've never prescribed it to anyone.

How does sacrifice work, psychodynamically?

I don't know. How does acupuncture work? I can't tell you, but it does. If you believe in it, you feel relieved. And don't forget that the native healers prescribing sacrifice are powerful, charismatic men.

Are there any women healers?

Very few, if any, although women in West Africa are members of the secret societies.

Is there greater incidence of mental illness among men or women?

Men. Why is this? They're in the front line facing the firing squad. They compete. They're under stress. They hunt. They kill each other.

Is this also true in the West?

It was until recently, but women are catching up fast. At one time, almost two-thirds of European women were taking Valium. After World War II, the rate of mental illness among women began to rise, but it's still not as high as for men, who are exposed to more dangers, crime, juvenile delinquency, social pathology, and prison sentences.

Is human sacrifice still practiced in Africa?

I am told that in the markets one can still buy human heads. There's no doubt human sacrifice was practiced as recently as ten years ago. Certain tribes in remote parts of Africa may still practice it. The oracle or some other voice tells you that the blood of a human must be sacrificed, otherwise the community will be wiped out by famine or another malevolent force. Men also kill to enhance their sense of maleness and potency. This resembles being thrown into the bush to fight lions as a test of manhood. If you come back alive, you are a big man.

Is this a manifestation of the castration complex?

No, the castration complex is not a physical state, according to Freud, but a mental attitude. What I'm talking about is actually physical, men making themselves feel important by beheading women and soaking themselves in blood as they chop each other up with knives and cutlasses. I'm not defending Africa, and this happens in other parts of the so-called Third World, as well, among aboriginal tribes in the mountains of Thailand, for example. But these practices are not generally talked about, so news of them doesn't surface.

I wrote a paper on a group called The Leopard Men Society of Nigeria, based on original studies by the African-American Stuart Cloete. At night the members of this secret society "changed into leopards" and committed ritual murders. They thought they would be immortalized by sucking out the blood of their victims. This was one of several major epidemics of violence in Africa. There were other occurrences of mass hysteria among the Odozi Obodo in Nigeria, the Poro Society in Sierra Leone, and the Mau Mau in Kenya. Something similar transpired with the myth of Mpaka-Fo. This engendered an acute castration anxiety that could only be warded off or expiated by tearing out the heart of a young child and offering it to Mpaka-Fo.

Is ritual murder part of juju?

Juju is a term that covers a hell of a lot of different things—you perform juju to marry a girl, put someone into a trance, send a supernatural message to your enemies, or kill someone. You could say the pope uses juju when he drinks the wine and eats the host. But generally it refers to people putting other people under a spell.

You coined the term "malignant anxiety." What does it mean?

It describes the psychic state of people like the Leopard Men, a condition of excruciating, impulsive anxiety that is action oriented. Once it has seized you in its grip, you *have* to do something about it—rip out the heart of an animal or kill someone. The phenomenon is similar to running amok in Southeast Asian cultures.

The person becomes deadly unstable and restless. He perspires tremendously. When you hold him down and ask, "What's wrong?" he replies, "I don't know! I don't know!" People say the spirits have possessed him. Men in this condition have been brought to me. I've given them sedatives and sent them home after they've slept for two days. But in other cases, when they have been taken to hospitals where the condition wasn't recognized, they have gone out and done serious injury.

Why do you call it* malignant *anxiety?

Because it's action oriented. You have to rush out and *do* something. The Leopard Men and other secret societies have developed in those parts of Africa that are being detribalized. When we open up villages to give them hydroelectric power and other modern developments, we are also opening up thousands of years of cultural history. In detribalized society you're on your own. You lose your social support, your sense of self. You suffer from depersonalization and derealization. You end up walking around the streets of Lagos and Ibadan disoriented, sleepless, feeling as if you don't belong. This is the penalty we pay for progress. It is a lesson we are learning again today in Eastern Europe.

Are the findings in your comparative study of mental health in New York City, Nova Scotia, and the Yoruba region of Western Nigeria still valid?

This was the first cross-cultural study of mental illness in the Third World. According to the missionaries, Africans had no inhibitions, mental problems, psychological or emotional disturbances. But we discovered schizophrenics and psychoneurotics who had been living in African villages for years. No society in the world is immune to mental illness, although some have the built-in community support that keeps it from manifesting itself. The European colonialists didn't understand any of this. So we exploded for the first time the myth of the "happy savage."

How does mental illness in Africa differ from that in New York City?

In Africa it's pure. If you're dealing with schizophrenia, it's schizophrenia pure and simple. In the West, schizophrenia will have multiple manifestations, and these manifestations will be masked. Africans recover more quickly and permanently than the other people we studied. This is because of the tremendous social support they receive in their villages. Illness in Africa is not individual, it's communal; psychotherapy is built into everyone's everyday interactions.

How did you decide that Western psychiatric methods wouldn't work in Africa?

I'm always a rebel. When I moved back to Nigeria after my schooling in England I was faced with two alternatives. Am I going to import the University of London into Nigeria, or am I going to start anew? I started anew, and I can tell you it was a tremendous success. I diverted from orthodoxy to develop a system relevant to my own people's culture.

Does Africa have forms of mental illness not found in Europe?

My colleagues thought the Yoruba tended to get manic without being depressive. But I told them this wasn't true. It was only from their Western perspective that the African looked manic. While the European is withdrawn, quiet, a cornered type of person, the African is so excitable that he's almost normally manic.

Is hysteria more prevalent in Africa?

Yes, but we use the term "hysteria" because there is no other word for it. In the evening an entire village will be crying and wailing as goats are killed for a burial ceremony, but by morning everyone is back to laughing and joking. Do we call this hysteria and talk of "dualism in the African personality," or is it really just tribal life? We need new words. Our psychiatric terminology is culture-bound, and the culture it's bound to is not our own. I prefer to describe these phenomena as "pseudo-hysteria." They resemble hysteria, but they are not hysteria. They are actually acute manifestations of grief.

Tell me more about pseudo-hysteria.

One day a young person was brought to me who had woken up shouting and screaming from a terrible dream. "Where am I?" she cried. "It feels like I'm still dreaming!" I gave her a shot of pentothal, and she went back to sleep for a day or two. The next week the mother came to thank me for curing her daughter. I'm like a magician, you know. In the old days she would have gone to the healer. The girl would have been tied down while the healer killed a goat or a chicken, and all the time he was performing his sacrificial rite, she would have been shaking and having bad dreams.

Do the healers sometimes harm their patients?

A lot of damage has been done, there is no doubt. They tie people down. Patients die of malnutrition or of diarrhea disease from bad water. That's why I thought it was so important to make their practice more sophisticated, because whether you like it or not, people still go to traditional healers rather than the hospital. The culture still believes this is the way to a cure.

What do you mean when you say that Africans display "herd solidarity?"

Sometimes whole villages would hire a truck and arrive outside the gates of the Aro hospital at four o'clock in the morning. One night I counted sixty-four people in three huge lorries. They might have been on the road for days, travelling up to seven hundred miles. The patient would be bound in ropes, suffering from schizophrenia or another form of psychosis. But if he were the son of a chief or the chief himself, the whole village would come in solidarity.

I took case histories from all these people, the patient's mother, his wives, cousins, uncles, and so on. How did his sickness first manifest itself? How long has he been acting strange? This wasn't just Mrs. Smith telling the psychiatrist about the odd behavior of her husband. This was the whole village recounting months of observations. Although I haven't been there for years, I understand people are still driving hundreds of miles from every part of Nigeria to get to "Lambo's hospital."

What are the drawbacks of tribalism?

It is supposed to be a balance wheel, but in practice it is a locked brake. Herd solidarity provides tremendous social support, but at the same time you have to obey its rules by not marrying outside the tribe and so on. Tribalism is a profoundly conservative influence.

Will tribalism survive in modern Africa?

It won't vanish, but it will be transformed. New social support mechanisms will take its place. Africa has many valuable things that the Western powers have lost. That's why I shout, "Don't throw out the baby with the bath water! Look to your own culture. Learn from it. Develop your own models for living."

Is the spirit world of the ancestors still a real force in Africa?

Tremendously so, and I hope it will go on for a long time. The ancestors support you. You go to their graves when faced with making an important decision. The mechanism is similar to confession in the Roman Catholic church. If you have nowhere to go, if you are alone in the world, you internalize your guilt, and the only way out is to commit suicide.

In Africa the atmosphere is still charged with supernaturalism. Even people who have got their Ph.D.s in England can fall back on their culture. When I was in Geneva, someone came into my office and said, "Tom, they're after me. So and so is using juju on me." This man wanted to get the Ministry of Health, and the other man didn't want him to. Within three days of becoming minister, the man was found dead in his chair, maybe because of a heart attack, maybe not. The other man was offered the chance to succeed him, but he would only take the job if the chair was destroyed. In Africa the gods are still alive.

Why did you decide to become a psychiatrist?

I wanted to be an anthropologist or sociologist, but in those days there were only two recognized disciplines for educated Africans—medicine and law. In medicine, the nearest thing to studying human behavior and social dynamics is psychiatry. It is the only medical discipline that looks at the whole psyche, the entirety of a person and his or her immediate relations. So I went to the seat of British psychiatry, the University of London's Institute of Psychiatry at Maudsley Hospital, where I was trained in psychotherapy and psychobiology. People had originally told me I was crazy to study psychiatry. "You're a good surgeon. You'll make a lot of money. They don't even practice psychiatry in Africa!" Well, today psychiatry in Nigeria is booming, and the universities are filled with professors of psychiatry who are my former students.

What are you working on now?

Other than running the Lambo Foundation, which supports young artists and works for social change throughout Africa, I am writing my autobiography. What in my nature allowed me to be so aggressive in pushing for ego dominance? I had my pipe dreams as a child, wanting to be a big person and solve some problems, but I was also in the right place at the right time. Since so many people believed in me, I had to respond to the challenge.

How far along is the book?

I'm almost done. I'm just validating my recollections by talking to people who are still alive, like my wife. I'm asking her what it was like dating a black African guy in Great Britain. Did she know at the time I was going to go so far, or did she just love me for who I was?

Wole Soyinka, a Nobel laureate in literature, comes from your hometown. Many Africans of your generation have excelled.

Yes, but the number who excelled is infinitesimal compared to the larger population. I have known people who worked even harder than I did, yet went down the drain. There must have been something inherent in us, something wanting to get out. Most of us who developed these leadership roles weren't trained for them. We assumed them spontaneously.

You once wrote that Westernized Africans are more anxious and insecure than their non-Westernized compatriots.

This is even truer today than when I wrote it. In colonial days, they decided whether or not you could go to Harvard, but they also guaranteed you a job. Now you can go to Harvard and still be a dropout. When I left Switzerland and went back to Nigeria, I asked my friends, "Where is so and so? 'He has disappeared.' Where is this other man? 'He is dead.'" Of the thirty-six people in my secondary class who went to university, only three of us amounted to something.

How did **you** ***manage to go from tribal Africa to postgraduate studies without becoming neurotic?***

It may be constitutional or due to my social background. People recognize me as a leader. They're always calling me to get Tom Lambo to chair something or other. So it pushes me. I have no time to be neurotic.

Beside psychotherapy, what other scientific knowledge is indigenous to Africa?

The Masai were suturing blood vessels, removing appendixes, and practicing other sophisticated surgical techniques long before the British. Without a vast herbal pharmacopoeia, most of Africa's tribes would long ago have been wiped out. Rather than merely imitating the West, Africa should build on its indigenous strengths. Innovate, don't imitate, I tell people, because Westerners themselves are unhappy with what they have.

So you don't believe in the technology transfer model of development?

The transfer of technology from First to Third World countries is a misnomer. It's a myth, a political stunt. No country will ever transfer its leading technology, because it wants to use you as a market. The Japanese only succeeded in transferring technology from the West because they went out and took it.

Africa today seems to be suffering from the Big Man syndrome—too many ruthless leaders with no idea of where they're going.

The major catastrophe at the moment is the political system. How can you influence these sorry bastards who are so motivated by greed? Those of us committed to the progress of Africa have sleepless nights. We feel lost, like Alice in Wonderland. For years, after every coup, I would meet with the new Nigerian president. I would spend hours telling him what's happening. But even when he meant well, the poor man was always surrounded by people who wouldn't let him do the right thing. Whatever progress he made was soon eclipsed by massive problems in health, agriculture, and the economy, which always ended up suffocating him under mountains of debt.

How can Africa avoid the psychological problems of development?

We have not prepared people for social change. For example, Nigeria woke up one day and said it wanted to build a big cement factory in a rural area. No one thought about the young men in the villages who would have to work there. None of them was trained to get up at seven o'clock in the morning. Time is timeless in Africa. After doing a two-year study financed by the Ford Foundation, I found the young men becoming progressively more confused, depressed, anxious. Absenteeism was climbing sky high. Building something overnight had caused tremendous psychoneuroses.

Very little could be done about the problem. I told the government and the bwanas in the factory, "Look, you've done the wrong thing. It's not enough just to clear the land and build a factory. You started this operation without giving the slightest thought to training the people who would have to work in it."

What model of development would you recommend for Africa?

I would hate for my country to be like Sweden, with its sky-high suicide rate, or like North America, with its teenage cocaine addicts. We must try as much as possible to avoid the mistakes of the so-called developed countries. Africa shouldn't compete with the Western world. We should retain our culture and solidarity. We need science and technology to guarantee the quality of life. We need improved educational and health systems. But we should also retain our spiritual dimension. In 1942, when I first went to England, somebody took me to church. There were two of us in this vast place shivering in the middle of winter. "What kind of utopia is this?" I asked myself. The East Germans must be thinking the same thing today.

Are you pessimistic or optimistic about the future?

Africans are resilient and courageous. Like a soccer ball kicked against the wall, they keep coming back at you. Europeans are fragile, while Africans are more agile, both physically and mentally. This is why we will absolutely survive. But you know, there is no culture in the world free of neuroses.

DORTOIR A
INFIRMERIE

Etienne-Emile Baulieu

✦ ✦ ✦

"I saw a lot of botched abortions when I was an intern. Abortion was illegal in France until 1966, but if a woman was in danger of bleeding to death, she could be brought to the hospital for a surgical scrape. I have horrible memories of those days." French medical doctor and biochemist Etienne-Emile Baulieu sat on the government committee that recommended France legalize abortion. More importantly, he developed RU 486, the "abortion pill." Today, more than a third of France's abortions are done chemically with RU 486. When this relatively safe and inexpensive antihormone reaches the world market, it will make abortion available to millions of women who today have no access to family planning.

"I want to help women," Baulieu says of RU 486, which he calls a contragestive, because it blocks the hormones required for gestation of a fertilized egg in the womb. "I have not dedicated my life to abortion. I am not antichildren. I have three children and seven grandchildren. But two hundred thousand women die every year in botched abortions, and RU 486 can save them."

A pioneer in research on hormone receptors and brain function, Baulieu struggled for twenty years to develop an abortion pill. When drug manufacturers were too frightened to investigate the idea, Baulieu used his position as consultant to Roussel-Uclaf, a $1.7 billion-a-year French pharmaceutical company, to nurture promising leads. After RU 486 was synthesized and its abortifacient properties accidently discovered by Roussel chemists in 1980, Baulieu mounted an even fiercer campaign to get the drug onto the market. He succeeded in 1988. The strings Baulieu pulled in maneuvering RU 486 through the top levels of the French government provide a plot worthy of Inspector Maigret. The cast of characters ranges from Sophia Loren, Baulieu's longtime friend and former mistress, to heads of the world's largest drug companies. Baulieu is plotting a similar coup for introducing RU 486 into the United States. "I love competition," he says, "and I like to keep score."

Baulieu claims he didn't even know what birth control was until Gregory Pincus, father of the contraceptive pill, recruited him in the early Sixties as a likely successor. Pincus spotted in Baulieu the traits that had made his own career so successful: the intellectual agility of a first-rate researcher combined with political flair. Still in his twenties, Baulieu had already isolated a water-soluble steroid from the adrenal glands of patients suffering from adrenal cancer. This surprising discovery cast new light on how hormones are transformed and transmitted in the body.

Appointed professor of chemistry at the University of Reims in 1956, Baulieu also kept a lab in Paris. Here he made his most famous scientific discovery in 1970, when he isolated the uterine progesterone receptor. Secreted by the corpus luteum, the outer lining of the ovulated egg, progesterone prepares the uterine wall for implantation with a fertilized egg, and without progesterone, the uterus cannot carry a pregnancy to term. Once the progesterone receptor was discovered, the next step in fertility control was to engineer an antiprogesterone, or antihormone, that blocks the uterine response to progesterone.

While looking for a chemical key to fit in his hormonal lock, Baulieu was appointed director of a laboratory at INSERM, the French institute for medical research. Since 1963, he has also worked as advisor to Roussel-Uclaf, where he had been offered, but declined, the job of research director. Searching for new ways to make cortisone analogs, Roussel chemist Georges Teutsch synthesized Roussel-Uclaf 38486—RU 486—in 1980. Baulieu shouted, "Eureka!" while Roussel's Catholic executives looked on in dismay.

After years of laboratory testing and political jousting, Baulieu finally arranged for the drug to be given to eleven volunteers seeking abortions. It triggered expulsion of the embryo in nine of the eleven women. Two had to undergo surgical abortions, and one woman required a blood transfusion. These risks of failure and excessive bleeding still exist, but the protocols for using the drug have since been improved. A single six hundred milligram dose, followed two days later by an injection or vaginal pessary of prostaglandin—a hormone that provokes contractions of the uterus—now has a ninety-six percent success rate.

After getting RU 486 into production and winning another fight to get the pill approved by the French Ministry of Health, Baulieu was dumbfounded when Roussel's president caved in to pressure from his superiors and pulled the drug off the market. Roussel is controlled by the German pharmaceutical company Hoechst AG. Hoechst's predecessor company was I.G. Farben, which manufactured the gas used in the Nazi concentration camps. When antiabortionists started comparing RU 486 to Farben's death-camp gas, the firm's president, an ardent Catholic who is sensitive to the American market, where he earns a quarter of his profits, halted production.

Shortly after hearing the news, Baulieu flew to Rio de Janeiro, where the World Congress of Gynecology and Obstetrics was meeting. Waving a petition signed by a thousand doctors, he denounced Roussel's decision as "morally scandalous." The French government owns about a third of Roussel, and by law it can withdraw a patent from a company refusing to use it in the public interest. Two days after Baulieu's press conference, the minister of health, calling RU 486 "the moral property of women, not just the property of the drug company," ordered Roussel to put it back on the market.

Roussel's policy since then has been to release RU 486 to countries whose ministries of health request it. This now includes France, Great Britain, Sweden, and Russia. The World Health Organization has an agreement with Roussel for distributing the drug in the Third World. In the United States, clinical trials are underway in Los Angeles, and Baulieu is confident the pill will soon be widely available, not only in abortion clinics, but also in doctors' offices, where abortion will be a personal decision between a woman and her family doctor. Deprived of specific locations to attack and the symbols on which its propaganda thrives, the American antiabortion movement should collapse—at least this is Baulieu's fond hope.

Baulieu lives with the style of an artist, although a friend says his life most nearly resembles that of P.T. Barnum. Induction into the French Academy of Sciences usually involves the ceremonial presentation of a sword. But in Baulieu's case, his friends Jean Tinguely and Nikki de Saint-Phalle arranged for him to be presented with a sculpture made by welding Baulieu's academic sword onto a chicken rotisserie. Baulieu works at a long table overflowing with papers, puppets, and masks. The walls of his office are lined with art and bookshelves holding the complete works of Pasteur, a set of *National Geographic*s, and the thirty-five volumes of Baulieu's bound publications. These papers document the invention of molecular endocrinology, a field Baulieu helped create in the Fifties when he started making his discoveries.

I spent an afternoon and the following morning interviewing Baulieu at the Bicêtre Hospital. This former madhouse and prison outside the southern gates of Paris is somber with espaliered trees and four centuries of ghosts. Baulieu's laboratory, specially built for him, occupies the second floor of a new building erected in one of the hospital's neoclassical courtyards. Here Baulieu the ringmaster orchestrates the work of sixty colleagues, including four secretaries and a librarian.

Baulieu's Gallic gestures fill the air, and a smile often hovers around the corners of his mouth. His office overlooks the hospital's family planning clinic, and as I later discovered from an informal survey, most of the women threading their way through his courtyard are on their way to RU 486-induced abortions, which are now safer and more private thanks to Baulieu's championship of women's rights.

✦ ✦ ✦

Why do you call the birth control pill "the most important invention of the twentieth century?"

By divorcing sexuality from reproduction, the Pill has revolutionized human behavior. For the first time in history it's possible to have a sexual relations without worrying about pregnancy. The Pill also demonstrated the ability of science to transform our lives. It's the forerunner to other chemicals, good or bad, that will alter brain function and behavior. The Pill has been elevated to mythic status. Among many thousands of drugs, it alone is called "the Pill."

But I'm astonished to see the same thing happening with RU 486. People ascribe to it either too much good or evil. It's either the work of angels or the devil. In truth, science has finally developed a drug for ending pregnancy, something that women have yearned for for thousands of years. I was on Italian television recently when the interview was interrupted with news of a Sicilian woman who had died from drinking an infusion of parsley, which is toxic in high doses; she was trying to induce an abortion. Even in Europe, there are still many cases like this.

At long last we have a safe way to interrupt gestation. But women are mistaken if they think they can take the pill under any circumstances without supervision of doctors or nurses. "Women are free at last!" they say. But this isn't true. I'm a supporter of women's liberation, and I think RU 486 is going to aid the cause. But I'm also a doctor concerned about the health of my patients. Five out of a thousand women have extra-uterine, or ectopic, pregnancies. RU 486 will not abort these pregnancies, which, if left untreated, are fatal. To demedicalize abortion by removing doctors from the process—it's insane! This is not a miracle pill; it's merely an instance of medical progress.

You've said the birth control pill "failed." What do you mean?

Out of the one billion women in the world of childbearing age, fifty million, or a mere five percent, take the Pill. The Pill has failed in the sense that something that is so effective and generally so well tolerated is used by such a small percentage of women. People grow tired of taking it month after month. Perhaps the minor side effects are one reason why women start and stop. It may not be the medical miracle we thought it was going to be, but it brought about the sexual revolution, and I still maintain that it's the most important medical discovery of the twentieth century.

Is RU 486 going to be less of a "failure" than the Pill?

Because the Pill, condom, diaphragm, and every other means of contraception have failed, RU 486 itself, unfortunately, will not be a failure. It will be widely used. I oppose sterilization because it's irreversible. I support reproductive freedom up to the end of one's reproductive life. But since all the remaining methods of contraception are imperfect, there will always be a need for RU 486.

You once said your character was predestined by your genes.

My father, Léon Blum, was a doctor and scientist in Strasbourg. He specialized in diabetes and was the first physician in Europe to treat patients with insulin. He had been widowed for ten years when he married my mother. She came from the "interior," as they say in Strasbourg, from Normandy and Paris, where she had worked as a lawyer. She gave up her career to have children, but my mother never got on with my father's family, and he probably made a mistake in marrying her. There were disputes, and by the time I was three and a half, he had died a young man.

When did you decide to follow in your father's footsteps?

For a long time I was hostile to the memory of my father, and my mother also was violently opposed to my studying medicine. She wanted me to become a Polytechnician with a beautiful uniform [the Ecole Polytechnique is France's military engineering school]. There was nothing pushing me in that direction, but one day I found myself simultaneously enrolled at the University of Paris in the faculty of sciences and the medical school. This cross between science and medicine is a hybrid. One is neither a pure researcher nor a typical physician. It used to be a strange thing to do, although today everybody understands the importance of linking basic research to medical studies. I put them together naturally, like my father, which is why I say in jest that my choice of careers was genetically determined.

How did your mother influence your career?

My mother was an exceptional woman, strong-willed, beautiful. She was an international lawyer before giving up her career to marry my father at age thirty-two. She was a very good pianist, with a master's degree in English from the Sorbonne. She adored England and greatly admired the suffragettes, the early British feminists. Believing in equality between men and women, she thought women should have the right to vote and the right to a job.

Was your family religious?

My father's family was Jewish. My mother's family must also have had Jewish antecedents, although she was raised without religion. As for me, I don't believe in God. I would *like* to believe in God. Like many scientists who know the extent of their ignorance, I want to know more. But I don't practice any religion, and the longer I live, the more surprised I am by the ferocity of religious disputes. People are remarkably attached to the *form* of their religion, be they Catholics or Moslems. But from my perspective, the only real question is whether or not one believes in God.

Do you think of yourself as Jewish?

I felt a certain closeness to the people when I visited Israel, but Jews to me are like long-lost cousins. They're part of my family, but far from my immediate world. I'm first of all a Frenchman. I love France, its history, the land. I'm also an internationalist, because science, after all, knows no national boundaries. Any scientist who contributes to the scientific enterprise is a friend.

Have you suffered anti-Semitism?

In school, when I still used the family name of Blum, I was called "dirty Jew" and things like that. As a young Communist, I saw anti-

Semitism in the Party. Between school and the Party came the anti-Semitism of Hitler's Germany. When my mother moved us to Grenoble during the war, we changed our name to Baulieu and managed to survive. The bourgeoisie in France, especially the upper echelons of the medical profession, has always had a strong vein of anti-Semitism. Whenever I confront these people, I call myself a Jew, even though I'm not, just to annoy them.

Does your being Jewish have anything to do with your discovery of RU 486?

Being Jewish has nothing to do with it. Fundamentalist Jews are indistinguishable from right-wing Protestants, Catholics, or Moslems. They're all fanatic moralists bent on limiting people's freedom. It's easier to work on abortion and contraception if one is part of the political left, where liberty and freedom of choice have always been important issues. My early interest in women's rights, other than my mother's influence, probably developed in the *auberges de jeunesse,* the leftist youth groups that were the precursors to the revolution of May '68. But I didn't start researching fertility control until I had already had three children!

Tell me about your involvement in the Resistance.

In 1942 after I'd gotten in trouble with the gestapo for breaking windows in the Grenoble militia, I went to Annecy and joined the Francs-Tireurs et Partisans Français, which were the irregular forces controlled by the Communists. But I wasn't a Communist then, just a fifteen-year-old kid still enrolled in high school. I was making up false identity cards for myself, my mother, and younger sisters when I chose the name Baulieu. I don't know why; I almost chose Baumont.

The FTPF got weapons from parachute drops. After the liberation of Annecy in 1944, I joined a battalion sent to the Alpine front. It was there I decided to become a doctor. Still wearing my uniform and the blue beret of a mountain infantryman, I came back on leave to take the medical school exams. They hurried things up for people in the army, so it was thanks to this brief military interlude that I had the leisure to start studying medicine and basic science at the same time.

Tell me about your postdoctoral research at King's College, London.

Max-Fernand Jayle, my advisor, sent me to England to study chromatographic techniques for looking at steroid hormones. He thought this would help him in his own research, but life is ironic, and it doesn't always turn out the way you want. In England I developed a taste for doing the kind of classic, analytical studies that Jayle himself couldn't stand. As a doctor doing basic research, I was able to make two or three early discoveries about the nature of steroid hormones. But this also precipitated my breakup with Jayle.

Tell me about your second trip abroad, when you met Gregory Pincus.

I didn't go to America to meet Gregory Pincus. I didn't even know what the word "contraception" meant. But whether or not you use the word, contraception is completely natural, and everybody does it. The French are known for early withdrawal. Like the rest of the world, they also use the rhythm method. The condom was never widely used in France, except by men having sexual relations with prostitutes.

Why did you go to the United States?

Seymour Lieberman wrote from Columbia University asking me to visit his laboratory, where he was doing excellent work on steroid hormones. The United States government denied me a visa because I'd been a Communist. It didn't matter that I had broken with the Party after the Hungarian uprising in 1956. I was talking earlier about the *auberges de jeunesse,* the libertarian side of the French left, but you also have to remember the strict morality of the French Communist Party, which was like a religion. The PCF was absolutely opposed to contraception and abortion. They held an internal trial to drum out of their ranks a very good gynecologist, Jean Dalsace, who had introduced the diaphragm into France.

What are your current relations with the Party?

I've never agreed with the people who quit the Party and became vehement anti-Communists. In my case, I just tiptoed away. Of course I'm anti-Communist, in the sense of opposing the loss of freedom, the gulags, the dictatorships, and failures. But among the rank and file there were some wonderful people. We aspired to a better world, with an idealism that was at once moral and scientific. It was totally false, but marvelous!

How did you finally get to the United States?

After Kennedy was elected president, I got a visa right away and went to work with Lieberman in New York, which is where I met Pincus. I'd actually shaken his hand once before, when he came to visit Jayle in Paris. But I was put off by his royal air. It was like shaking hands with God. I was initially disappointed that Pincus, although a very good scientist, was doing more politics than science. I didn't realize at the time that he was making a revolution.

When he invited me to give a talk at his endocrinological laboratory, I didn't go because of the Pill, which I knew little about. I went because he was a famous man who might be able to launch my career. It's thanks to Pincus that I became part of the network of people involved in fertility control. He flew me to Puerto Rico to visit the laboratory where the Pill was being developed. He turned out to be a very persuasive man, but I never promised Pincus I was going to work on the Pill. It was not reproductive biology that interested me, but the mechanism of hormone activity. I wanted to look at sex steroids from the biochemical viewpoint.

How did he convince you to work on contraceptives?

I was a thirty-five-year-old lab scientist, and you can't be a practicing physician while you're trying to do research at the highest level. But research on contraception wasn't only molecules and cells. It was applicable to people's lives, and there weren't many scientists engaged in this kind of work, at least in a serious way. Everyone has an ideological position on contraception, but reproductive biology is still largely unexplored terrain. I was also influenced by the idea—which was then being taken very seriously in the United States—of overpopulation in developing countries.

World population will double sometime in the twenty-first century, from five to ten billion people. This will cause tremendous problems and disparities in certain regions. There will be twice as many Moslems, while the populations of Europe and North America will remain the same. A world with ten billion inhabitants is livable, but RU 486 and other methods of birth control will be essential for maintaining a stable population. Without birth control, there will be no social or technological progress. It is also a woman's right to decide how many children she wants.

Was the birth control movement active in France?

All forms of contraception had been outlawed in 1920, when the country was desperate to repopulate itself after the war. There was the famous case of a woman in Vichy France who had her head chopped off for performing abortions. Diaphragms, spermicides, the Pill—all were illegal. Only condoms were tolerated as a way of preventing prostitutes from spreading venereal disease. Everybody ignored these laws, but France was not sociologically prepared for legalized contraception until Mitterand, the presidential challenger, made it an election issue in 1965. De Gaulle set up a committee to study the matter, thirteen wise men, and I was one of them. This was my first small intervention in French history!

Did Pincus help your career in other ways?

He arranged for me to sit on a World Health Organization committee, which was a clever maneuver to have me meet demographers, gynecologists, social scientists, and other people working on the Pill. Pincus thought this would enlarge my vision and make me more inclined to contribute to fertility control. And he was right. I became a defender of the cause.

Thanks to Pincus and Lieberman I was invited to give a presentation on estrogen receptors in the rat endometrium at the Ford Foundation in 1965. The Ford people were stupefied. I was developing work done earlier by the University of Chicago chemist Elwood Jensen, who discovered the estrogen receptor. To me it was obvious that Jensen's approach was the way of the future. Hormones travel through the blood to cells that respond to them—uterine cells, pituitary cells, whatever. The receptor is the specific locus where the hormone hits the cell and tells it what to do. This is the likely place to intervene. I was applying to hormone research what pharmacologists had been saying—but not doing—for fifty years. Already in the early 1900s German chemists were picturing receptors as locks with hormonal keys. But no one had yet found a way to open and close the door.

What happened after your talk at the Ford Foundation?

I told them I'd work on sex hormones and their receptors, but I wasn't promising to invent a new pill. I told them to be confident that sooner or later something would come out of basic science. They were nice and clever enough to give me money, enough money to last for ten years. So we discovered the progesterone receptor in the uterus and the androgen receptor in the prostate. We introduced the concept that endocrinology involves changes both in hormones and receptor concentrations. For example, at the beginning of a woman's cycle she secretes estrogen, and then in the second part, she secretes progesterone. Estrogen triggers ovulation, but it also induces the synthesis of the progesterone receptor—"primes" it. This is an important discovery from my lab. We also found that progesterone deactivates, or "down regulates," its own receptor before the onset of menstruation. These mechanisms apply to all hormones.

Could understanding these mechanisms lead to the development of a new form of birth control?

A steroid that binds to the progesterone receptor could provoke this down regulation of the receptor. Although theoretically a very good contraceptive, it would have to be given to a woman exactly at midcycle. But what is midcycle? Even regular cycles are always to some extent irregular. Nonetheless, some people in England and Stockholm are experimenting with RU 486 as a midcycle contraceptive. One would take antiprogesterone very early, just after ovulation, so that the endometrium is unable to implant a fertilized egg.

When did you begin working on the progesterone receptor?

Already by 1964 I was letting my earlier work on adrenal metabolism die out in order to work exclusively on receptors. To create the new field of molecular endocrinology, I saw that the trick was to isolate the receptors. This originally had nothing to do with fertility control. I wanted to understand the general principles of hormone action. I could have worked on any number of hormones, and even among the steroids, I had a choice between corticosteroids and sex hormones. They're all the same for my purposes, which is understanding the mechanisms of molecular function, but I deliberately chose to work on sex hormones.

Why?

Because of fertility control and because I'd been seduced by Pincus's way of thinking. Work on sex steroids would involve me in an interesting social activity, which I had lost when I left the Communist Party. It was also a way for me as a research scientist to get back in touch, indirectly, with medicine.

What happened after you discovered the progesterone receptor?

Roussel and other drug companies pulled out all the molecules in their drawers to look for antagonists to progesterone. But we hit a dead end, so I played some more with the idea of receptor regulation. I was building on Arpad Csapo's work in St. Louis. He was the first to prove that progesterone is indispensable in women. That work made me confident that an antiprogesterone would be effective in blocking pregnancy. And it's very important in science to be confident.

Georges Teutsch, the chemist who synthesized RU 486 in 1980, claims to have made his discovery independently of your own ideas.

Teutsch was part of a group of chemists working on corticosteroids, whose antiinflammatory and antiallergic effects earn drug companies much more money than sex hormone drugs. Our two lines of research merged at a certain point by chance. I'd been presenting my ideas on antihormones to the R & D committee at Roussel. As a Catholic, conservative company, the directors were opposed to working on sex hormones. They preferred to work on corticosteroids rather than progesterone, but the two have a lot in common, and I knew that antihormones for one could be applied to the other. As in any hierarchy, there are gaps. I was talking to people at the top, the chief of pharmacology, and so on. But chemists down below were doing their own work, and they suddenly came up with eleven substituted molecules—just the type of structure I had suggested on the basis of our own work with other antihormones, such as antiestrogens.

What happened after Teutsch synthesized RU 486?

We had to follow the rules of the high committee, and since it was officially forbidden to work on antiprogestins, we had to test RU 486 as an antiglucocorticosteroid—which includes all the drugs that counteract the activity of cortisone and cortisonelike hormones. RU 486 was synthesized, patented, and tested as an antigluc, but by a simple test in animals, it was also found to be an antiprogesterone. I was really keen on having a drug that induced menses, as well as an abortifacient. So the next step was to test it in humans, using women who were both pregnant and not pregnant.

Didn't you first have to do more extensive tests in animals?

Yes, it was sent out for toxicological studies in monkeys, who were given a hundred milligrams per kilo every day for a month. An enormous dose. Three monkeys got sick and had to be killed. The toxicological people at Roussel who received the report told me nothing, until one day at a committee meeting, somebody said to me, "By the way, your compound is dead." Which meant it was too toxic for human use.

I looked at the studies and found RU 486 doing exactly what an antiglucocorticosteroid should do: block cortisone activity and induce extreme weakness, hypotension, kidney failure, and other metabolic disorders. "This is beautiful," I said. "The compound is working in vivo!" So I rescued RU 486 from oblivion. I put some in my pocket and went to Walter Herrmann in Geneva, where we did the first human tests. People were initially afraid that what happened to the monkeys would happen to the women. They didn't understand that one small dose given for the regulation of progesterone is completely different from using the drug as an anticorticosteroid for many successive days at high doses.

What was Roussel's reaction to the human trials in Geneva?

People were extremely excited, but Roussel's president, Edouard Sakiz, was not confident he could find the money to pay for real trials. That's when we arranged for the World Health Organization to make the trials, because the drug is potentially of great interest to the Third World. When the WHO tests confirmed our first experiment, I and a number of allies inside the company convinced Roussel to pay for trials in France.

After your big fight with Roussel, are you still working with the company?

Edouard Sakiz has been a friend of mine for thirty years, since we met in graduate school. I introduced him to Jean-Claude Roussel in 1965, who hired him as head of a lab, a position relatively low down the line. Sakiz climbed up to become president, and we are still real friends, which is quite an achievement. We had breakfast together yesterday in the Roussel dining room. Actually, Sakiz and I are on better terms than ever, and I'm still consulting for the company. RU 486 is ultimately going to be released around the world, but yesterday morning we discussed the U.S. strategy, which we agree on. First, we're going to make a study—

political, social, economic—which will probably conclude the wind is blowing in our favor. Scientists are clearly for RU 486, and even the Republican Party is going to change its mind. The women's movement is even stronger since the *Webster* decision from the Supreme Court.

Secondly, Sakiz has decided that Roussel will not hide behind its finger, as we say in French. Roussel will go ahead in America by itself. This will be difficult since Hoechst, the parent company, is not ready to do it, nor does Hoechst have expertise in the field of obstetrics. So Roussel has to find partners. These can be other companies, but I predict big companies won't touch it. So that leaves smaller ones. Another possibility is venture capital. We've had at least ten serious offers to create new companies, and a couple are extremely interesting. The third possibility is to ally ourselves with nonprofit groups who already have a network of centers that could deliver the pill. I am suggesting that Roussel, venture capitalists, and Planned Parenthood form a coalition to create a single-product company.

What about Federal Drug Administration approval?

The FDA is not a problem. They'll do their job quickly and well, because they're good scientists. They already know the drug works from the published data. My overall strategy from the beginning has been to work in the public eye. This is based on my confidence in science. If the facts are good, people buy them. The studies already done are convincing for any scientist of good faith anywhere in the world. More than sixty-five thousand women in France have taken the abortion pill. We're up to one thousand a week, and each week there are more than before. One third of the abortions in France are now done with RU 486, and we're still on the curve.

Will all abortions ultimately be done with RU 486?

I don't think so. Some women won't like it. With a surgical abortion, you close your eyes and put yourself in the hands of the doctor. RU 486 involves a more active approach. You take the pill yourself, and the process lasts longer, since you have to wait forty-eight hours before the bleeding starts.

What other countries are going to start using RU 486?

It will go to England immediately. Then comes Sweden and the Scandinavian countries. And very soon afterwards, Italy and Spain. Abortion was legalized only two or three years ago in Spain, but we've had a few trials, and the press is very much in favor. When we see what happens in England, France, Italy, and Spain, then we'll work on America. Russia has just asked officially for RU 486. As for moving into the Third World, there is no "no." In fact, this was one of my motives for developing the pill. But releasing it in the Third World is not the first priority, for practical reasons, and it will probably have to be undertaken by an organization that specializes in implanting new medical technology in developing countries.

China registered RU 486 in 1988, three or four days before France. They sent me a telex asking to buy enough pills to distribute free in the country's eight best medical centers. Naturally, I was very happy. I handed the telex to Roussel, and they never replied. This was during the retreat, when the policy was "no" for distribution of RU 486. Money was not the problem, because Roussel can make money on the pill. The problem was political pressure. Now there's been a reversal, but Roussel is waiting for an official request from China, because their policy is to answer only official demands. It was the same in France, where the minister of health had to order them, "Do it."

Will the drug be used differently in different cultures?

The best way to use RU 486 anywhere is to have women who have missed their periods go as soon as possible to a specialized center for the induction of menses, contragestion, or whatever you want to call it. Many women who miss their periods wait another month before doing anything. But they shouldn't. If they decide to have an abortion, they should act as soon as possible. French law says you have to have an abortion within the first three months, and I think that's a good time to fix the limit.

Will there be more abortions because of RU 486?

Not more but simpler, safer, more private abortions. When it comes to fertility control, I've always said, "The earlier, the better." Suspension of ovulation and contraceptives are better than treating the problem later. But many women don't use contraceptives, and many contraceptive methods fail. So even in an ideal world, you'll still have abortions. RU 486 will never become *the* method replacing other contraceptives, although this is a theoretical possibility. Without birth control a woman with a normal sex life will be pregnant two or three times a year. She sees her period is late, oops, she takes RU 486. This is not what I'm recommending now. Maybe in twenty years.

If RU 486 is safe, why not use it two or three times a year?

Women want real freedom, not just words. To tell them, "Don't bother with precautions. You can do anything you want. If you have an accident, RU 486 is here, so it's alright"—this is playing with people. We don't know the long-term effects of repeated interruption with RU 486. I want the knowledge brought by progressive experience. However, there are certain cases where RU 486 might become the sole method of contraception. Take the forty-year-old woman living alone who doesn't use birth control pills because they're bad for her health, or a young woman who doesn't have a sex life except occasionally, and then something happens. In these cases RU 486 could be very useful. Out of the 1.6 million abortions performed every year in the Unites States, more than a quarter involve teenagers.

Could RU 486 be used as a once-a-month or postcoital birth control pill?

It could be used once or several days a month as a contraceptive in the classical sense of the word, that is, to block ovulation during the first part of the cycle. It would be an estrogen-free contraceptive functioning like the Pill. It might function *better* than the Pill, because estrogen is what gives the Pill its harmful side effects. But who's going to develop this new method? It's easier and less costly for drug companies to play with what they already know about.

We've suspended ovulation in monkeys for six months with a single injection. But human trials would be very expensive, and we have no evidence the drug will be tolerated if taken regularly. So I doubt it will ever be clinically tested. RU 486 will remain a pill of the occasion. It could also be used once in awhile as a postcoital contraceptive, a day-after pill. In case of sexual exposure at midcycle, it can be taken late in the luteal, or second, phase to induce abortion. It works ninety-six percent of the time. But that's not efficient enough to use each month.

What other hormones are you currently investigating?

Our most interesting and unexpected discovery concerns the interaction of steroid receptors with something called "heat shock protein." If you heat up cells, they shut down their functions, but before dying they synthesize specific heat-resistant proteins. These molecules are conserved over evolutionary history. You find them in every organism from bacteria to humans. We've cloned one of these proteins and are studying its involvement in steroid receptors.

We're working on other steroid interactions at the gene level. These hormones stop or start the synthesis of many proteins, so there are many systems to study. We've also discovered that the brain makes neurosteroids that affect neurotransmitter function. We're studying the effects of hormones on aggressive behavior. We're also looking at aging of the brain, multiple sclerosis, and other degenerative diseases. This research is at a very primitive stage, but I hope to develop general systems that apply across species, including humans. I like to unveil principles, but I am a practical man. I am not so much a theoretician as a doer.

Are you a feminist?

Yes. As the poet Louis Aragon said, "Woman is the future of man." I'm for absolute equality before the law, but I love difference, and the simplest difference we know is between men and women. Women's reproductive capacity gives them a lot of special problems. Their brains are different; everything about them is different. But the rights of men and women should be the same. Women's liberation will go forward, but we have to keep pushing. We're going to see more women doctors, scientists, heads of state. But I doubt women will ever play tennis better than men.

What about the rise of religious fanaticism and right-wing pressure groups?

Obviously these endanger the movement. But I have tremendous confidence in the power of women. Women will always win in the end because they make the children. The caress at the beginning of life is a determinant factor. Men will never overcome the bond between mother and child.

Are you ambitious?

To be a scientist is to be ambitious, otherwise one doesn't do it. But I'm talking about ambition in the American sense of the word. I want to make discoveries, advance the field. In France, to be ambitious is to yearn for power and money. You have to be an idiot if you think science will reward this kind of ambition. It may be stupid, but nonetheless, scientists, like painters, writers, and politicians, succumb to the desire for fame.

This contradicts the scientific ethic and the practice of science, but still we want to be known for what we have done.

A serious problem of modern civilization is people's ignorance of science. They don't understand it and are afraid of it. Even the ecology movement, which I strongly support, has an irrational element that blames the world's problems on science. Nobody cares about the lives of scientists. The scientific enterprise is depersonalized. Science is the science of nobody. It's the science of robots, computers, and men in white coats. It produces miracles, but no one understands how. You can pick up a newspaper and read about Dr. Smith's discovery of a new drug for treating AIDS. And what will you remember about this article? Not the name of Dr. Smith. It's exactly the opposite for an artist like Andy Warhol. You can easily forget the subject of his paintings, but you won't forget his name.

Aren't you more famous than most scientists?

I was a scientist like all the others, known among my peers for a few small discoveries, until some accidental circumstances led me into an area where science and society encounter each other. The same thing happened to my colleague, Luc Montagnier, the discoverer of the AIDS virus. All of a sudden, a man in a white coat was mixed up with sex and drugs. But this rarely happens, and even without this direct bearing on social problems, I would still like scientists to be better known—even if I can't personally complain!

I'm not a nineteenth-century moralist, saying that science works only for the good of humankind. But I do believe science is the most important element in the evolution of modern life. When I descended into the grottoes at Lascaux to see the prehistoric cave paintings, I had a revelation. These superb images felt entirely modern, and to my amazement I realized that over the centuries humans haven't changed at all artistically. But scientifically, thanks to the interaction of our brains with nature, we've changed a great deal.

Tell me about your book,* Génération Pilule, *The Pill Generation. Is it like James Watson's* Double Helix, *which I know you greatly admire?

I'm not as talented as Watson, and I have a lot more enemies! My book includes a pedestrian explanation of how RU 486 works, the nature of its discovery, and a little bit about my life. I want it to be sold in supermarkets and bought by women out doing their shopping. One chapter will be a kind of user's manual.

Have you ever thought about writing a novel?

It's an idiotic dream. In that domain I'd be *too* ambitious. It's Flaubert or nothing. It would be better for me to try to make one or two more small discoveries in biology. Finally, in one fashion or another, I think only of my work. I sometimes have my best ideas when I'm out with someone or flying in an airplane. To get anywhere in this world, one has to be obsessed.

Richard Dawkins

✦ ✦ ✦

NOTHING in his scientific training or twenty years' experience with computers prepared Richard Dawkins for what emerged on his screen one night. The Oxford zoologist had designed a program called *The Blind Watchmaker,* which simulated the process of evolution by mutating electronic "genes" or quasibiological forms he called biomorphs. "I began to breed generation after generation of biomorphs from whichever child looked most like an insect," he recalls. "As I watched these exquisite creatures emerging before my eyes, I distinctly heard the triumphal opening chords of *Also sprach Zarathustra* in my mind. That night 'my' insects swarmed behind my eyelids as I tried to sleep."

Ever since he published *The Selfish Gene* in 1976, Dawkins has asserted a gene's-eye view of the world and its workings. "Humans are nothing but temporary survival machines, robot vehicles blindly programmed for someone else's benefit," he says. The true rulers of the world are the bits of DNA that make up our genes. Dawkins compares these genes to successful Chicago gangsters. Having survived for millions of years, they are the only immortal part of the body, and the secret to their success is "ruthless selfishness."

Thousands of Dawkins' readers have written to tell him *The Selfish Gene* changed their lives. It made them understand for the first time that genes are the building blocks on which evolution works, the key to understanding animal behavior, the stuff of life itself.

In his second book Dawkins carried his "fundamental law of gene selfishness" a step further. *The Extended Phenotype,* published in 1982, shows how genes manipulate not only the bodies in which they happen to be sitting, but also those of other organisms. Parasites hijack their hosts' tissue. Male mice subjugate females by emitting pheromones that drive them into heat. One can even say that "sneezing genes"—commonly known as viruses—have mastered the knack of replicating by getting themselves passed from nose to nose. Competing gangs of genes, claims Dawkins, "are running arms races in evolutionary time," the stakes being reproductive immortality or extinction.

Lest Dawkins be mistaken for a genetic determinist, *The Selfish Gene* ends with a call to "rebel against the tyranny of the selfish replicators" that are our genes. Dawkins posits another unit of evolution, a nongenetic replicator, the *meme.* Memes are ideas, bits of consciousness, beliefs capable of evolving, combining, and flowing down the generations. Our ability to choose between replicating genes or memes is what distinguishes humans from other animals.

If humans have broken free from the genes that program them, then such a change could happen again. Only this time, Dawkins says, it will be conscious machines—computers—that take over from humans to replicate their own forms of life. These and other speculations on the evolution of complex systems are the subject of his third book, *The Blind Watchmaker,* published in 1986.

Dawkins does most of his thinking these days sitting in front of a Mac II computer running his *Blind Watchmaker* program. Biomorphs, he feels, are a uniquely powerful and flexible tool for exploring his current "heretical" speculations on embryology and the origins and development of life itself. His method involves "switching back and forth between real life and the computer." Our conversation followed the same format. The discussion in Dawkins's Oxford apartment was interspersed with sessions on the computer touring Biomorphland.

The son of an agronomist in the British colonial service, Clinton Richard Dawkins was born in 1941 in Nairobi, Kenya. After inheriting a farm in Oxfordshire, his parents returned to England to raise dairy cattle. Growing up within twenty miles of Oxford and eventually going to school and teaching there, Dawkins took his father's interest in natural history and gave it a theoretical spin. This, combined with a brilliant prose style, have made him a best-selling author and recognized master at explaining evolutionary process.

Apart from two years' teaching at the University of California at Berkeley, Dawkins' entire professional life—as an undergraduate and graduate student in zoology, as a research assistant to Nobel Prize-winning ethologist Nikolaas Tinbergen, and as a university lecturer in animal behavior, a post he has held since 1970—has revolved around Oxford University.

The twice-divorced father of a young daughter, Dawkins lives alone on the top floor of a house with windows overlooking the medieval towers of Oxford. In this airy attic with the casual look of student digs, his office chair is positioned in front of a cockpit of video displays and computer keyboards. A slender man with a boyish face and schoolboy mop of hair falling over his ears, Dawkins speaks as eloquently as he writes. And the subject on which he loves to discourse is life—how it evolved and where it is going next.

✦ ✦ ✦

How did you develop your gene's-eye view of the world?

It's inherent in Darwinism. The body has within it a sequestered, privileged subset of cells: the genes contained in our sperms and eggs, which are the only parts of the body that are immortal. The German biologist August Weissman, at the turn of the century, first verbalized this separation between what he called the germ line and the body. I've just carried this implicit line of thought to its logical conclusion and turned it into a radical metaphor that caught people's imagination. Perhaps it's not a metaphor; in some sense it's the literal truth.

What are genes and why do you call them immortal?

A gene is a length of DNA, but for my purpose, we could have said what's important about genes long before DNA was discovered. Genes are passed down the generations. They have the power to go on in the form of exact copies of themselves for millions of years. This is why I think the body is a temporary survival machine, a vehicle for the genes that ride inside it. The fate of the genes in their quest for immortality is bound up in the short-term success of the body they inhabit, or the long succession of bodies they inhabit, because they go from body to body. Successful genes are those that make a long succession of bodies good at passing them on, which means good at surviving and reproducing.

Why do you call genes selfish?

Genes take whatever steps are necessary to survive. If an animal is caring for its young, this may be altruistic behavior at the level of the individual organism. But genes are controlling this behavior, and the genes in this case have been copied in the body of the offspring being cared for. *All* examples of apparent altruism at the individual level are the result of selfishness at the gene level.

When did you begin thinking that genes ran the world?

I was doing postdoctoral studies with Nikolaas Tinbergen in 1966 when he asked me to give some lectures. This got me thinking about why animals behave the way they do. The best way of conveying these ideas was to talk about the genes as being in control of life. The rhetoric of immortal genes leaping down the generations, jumping from one throw-away survival machine to another—it's all in my Oxford lecture notes. It was unconventional imagery to get across essentially orthodox ideas.

***Was that what you hoped to do in* The Selfish Gene?**

When I started writing it in 1972, *The Selfish Gene* was an attempt to get rid of the group-selectionist view. This outright wrong idea had obtained a grip on the popular presentation of science. Time after time I'd see excellent natural history programs on television marred by this false assumption that individuals act for the good of the species or the good of the ecosystem or the good of the world! This was an error that needed exploding, and the best way to demonstrate what's wrong with it, I felt, was to explain evolution from the point of view of the gene.

Overnight* The Selfish Gene *became part of our daily reality.

Perhaps that's because it brings home to people the truth about why they exist, something they previously took for granted. No one had given them such a ruthless, starkly mechanistic, almost pointless answer. "You are for nothing. You are here to propagate your selfish genes. There is no higher purpose to life." One man said he didn't sleep for three nights after reading *The Selfish Gene*. He felt that the whole of his life had become empty, and the universe no longer had a point. Another way of putting it is to think of people losing their religious faith. Where previously they had been fobbed off with religious, pseudo answers, they now felt they understood what life was all about. Though it sounds like a negative message, it has had a great impact on people.

I personally don't find the book pessimistic. In the end you talk about our brains giving us the option of escaping the tyranny of our genes.

Brains are part of the machinery for propagating selfish genes. But in the course of becoming more efficient at effective gene survival, they've acquired the capacity to rebel, to take off in their own directions. This is particularly true in cultural environments, where whole populations of brains can get together and communicate. A social federation of brains can take evolution in directions radically different from those favoring the replication of genes.

Opposing genes, you posit memes, or cultural units of reproduction.

We can lead fulfilled lives reproducing ideas rather than offspring. I'm interested in the possibility of ideas taking on a life of their own. I'm intrigued by the prospect of self-replicating entities that have the power, by mutation and selection, to club together into more and more efficient units of self-preservation. In *The Selfish Gene* I use the example of religions, which I see as collections of mutually compatible memes, just as bodies are collections of mutually compatible genes. When you've had a large number of generations for selection to go on, it becomes quite plausible that sets of mutually compatible memes will get together and become the great religions of the world. The same is true for collections of political memes, but I think religion provides the best example, because it is ultimately futile, and I really do want to emphasize meme survival for its own sake.

What is the likelihood of our finding self-replicating entities, either memes or genes, elsewhere in the universe?

I agree with the modern conventional view that life is widespread in the universe as a whole, but probably not so widespread that we are ever likely to meet it, which is a pity. But in *The Blind Watchmaker* I advance the less orthodox view that we are the only form of life to be found anywhere in the universe. When we think about what had to happen for life to originate in the primeval soup, from one perspective it looks very improbable, so improbable that it might have happened only once in a universe of billions of worlds. But I am actually interested in what I think is the more likely scenario that there is lots of life in the universe.

Could life be defined as something that replicates itself?

Something replicating itself is the necessary prerequisite for Darwinism to get going, and I believe that Darwinism is the prerequisite for all forms of life. Although self-replication is the prerequisite for life, life doesn't immediately follow from self-replication. Life follows from many, many generations of gradual evolution after self-replication has begun.

Will the self-replicating entities discovered elsewhere in the universe be different from life as we know it?

Such self-replicating entities could be some molecule other than DNA. Perhaps they could even *be* DNA molecules, but they almost certainly won't use the same genetic code that we do. Plenty of other molecules may be capable of entering into a self-replicating system the way DNA does. But these self-replicating entities might not even be molecules. They could be macroscopic entities, things big enough to be seen with the naked eye, such as cloud patterns. Anything that's self-replicating could be the basis for life.

Could we look for life elsewhere in the universe and fail to recognize it?

We could easily look at the self-replicating *entity* and not recognize it. But I doubt that we could look at the manifestation of a hundred million generations of evolution and not recognize it. And if we didn't recognize it, then it wouldn't be a very interesting form of life anyway. I'm inclined to think that life began with some other kind of self-replicating entity, and this provided an environment in which DNA was able to take over. Whether it's the original replicator or a takeover, DNA is very good at what it does. It's extremely faithful as a reproducer. The mutation rates are around one in a million. It's even evolved elaborate proofreading systems to remove most of its mistakes. Fortunately it doesn't remove all of them, or evolution would come to a halt.

If DNA weren't the first replicator, could something succeed it?

Yes. Our brains have provided a silicon milieu in which some kind of further takeover is possible. An electronic takeover would seem to be the most likely. I am fascinated by complex functioning as a diagnostic feature of life. If you went to another planet, how might you recognize life? If it were something like a computer, you'd see immediately that such a complicated piece of organization was created for a purpose. So you'd know life was or had been there. Life, I believe, can come about only through a Darwinian process of natural selection. Whether artifacts themselves are naturally selected is problematic, but they wouldn't be there if it were not for life. Computers are a particularly interesting form of artifact, because they provide a medium in which all sorts of lifelike processes can go on. Computers evolve in the sense that computer designs—the ideas for how to make computers—jump from brain to brain via language and written plans. You can make an evolutionary succession of computers that improve as the years go by. There is a kind of germ line, the equivalent of genes, in the blueprints for making computers. Computers themselves don't breed; they are used for awhile and then tossed in the scrap heap. But the ideas that went into making them can breed as genes do.

When are we likely to see self-replicating machines?

Self-replicating hardware may not happen, because there's no obvious reason why humans should want it to happen. Self-replicating software, on the other hand, already happens, with people copying files and sending them off to each other. This self-replication at the software level has sinister implications. You could have computer viruses capable of mutating in such a way that they become more invasive.

Could you have self-replicating hardware?

A computer could easily have a copy of its own blueprint in its memory. Then all it would need to build another computer would be the equivalent of limbs—robot arms that would reach out, pick up integrated circuits, and plug them into their sockets. A factory making computers could be automated to run without human intervention. Whether these machines could launch a sinister escalation of evolution depends on the amount of random variation they allow in their assembly plan, their program.

But I don't think a takeover by computers of the human world will come through computers manufacturing hardware copies of themselves. Robot arms are terribly inefficient compared with human arms for doing this kind of intricate work. More likely, the computing systems that control government organizations, like the Pentagon, could become more and more autonomous from human decision making. When artificial intelligence becomes more sophisticated, I don't see why large organizations, where humans now do the donkey work, shouldn't be taken over by machine intelligence. Humans will then sink into the background as functionaries or slaves.

When will this happen?

I don't see it happening in the near future, but that's because I'm a child of my time, who doesn't have this sort of vision. But if you extrapolate from the way computers have evolved during the last thirty years, we could be in for a nasty shock. For all I know, the enslavement of humans by machines could be a positive thing. But I don't find it an attractive possibility, aesthetically.

Have you seen examples of artificial intelligence that are truly intelligent?

Since I believe that I'm a physical machine, I can't escape the conclusion that it would be possible to build a computer that had the same kind of consciousness as I have. Whether computers will evolve consciousness by some selective process or acquire it by design, I don't know. The way we design computers today could be entirely the wrong track for getting to consciousness.

What is required for the evolution of consciousness, and why did it happen in humans?

It's extremely puzzling. An efficient survival machine in a very complicated environment would benefit from having an onboard computer enabling it to behave in complicated ways. The more complicated the behavior the better, because it can respond to a wider variety of contingencies thrown at it by the environment. Beyond a certain level of complexity, the most efficient way for a survival machine to organize its behavior is to run a computer simulation of the world inside its head. That's what we do when we imagine things. Simulation programs are now commonplace in this computer world, and none of them is conscious. They don't need to be. So I don't know what consciousness is good for that could not be done by highly complicated, but unconscious, simulation.

Have you any ideas where consciousness came from?

In highly social animals the most difficult part of the environment involves outwitting fellow species members, courting, fighting, or competing with them. Second guessing what other individuals are likely to do becomes a major part of surviving.

Another theory comes from the philosopher Daniel Dennett at Tufts University, who uses a computer analogy. There are basically two different kinds of computers: parallel and serial processors. My Macintosh is a serial processor. It has a single computing engine that deals with problems one at a time, even though it does this exceedingly fast. Our brain, on the other hand, looks like a parallel processor. We do many things all at once. But our subjective stream of consciousness is not parallel. It presents one serial phenomenon after another. Dennett thinks that consciousness is a serial machine running atop a basically parallel computer.

Do you have only one child because you're more interested in memes than genes?

I don't see a strong link between my personal life—my decision to have a child or not—and my intellectual life. I think these things happen the way they happen without regard, in my case, to the fact that I'm interested in the propagation of memes. If things had been different in various ways, I would have liked to have more children. But I didn't have a burning desire to have lots of children, and I still don't. I am delighted with the one I've got.

Birth control provides one of the best examples of how humans have revolted against their genes.

There are no contraceptives in nature. All the selfish genes need to program into us is sexual appetites. Then we have no control over the rest. But now that we have artificial means of preventing conception, we can enjoy our sexual appetites and yet have no children. We desire children for their own sake, but it's not surprising that so many people happily subvert the designs of their genes by contraception.

Tell me about your family.

My mother went to art school, but didn't do professional work after she got married. My father read botany at Oxford and then did a further degree in tropical agriculture before going to Africa, where he worked as an agricultural educator. We moved back to England when I was eight, because my father had been left a family farm in Oxfordshire by a distant cousin whom he never knew. He gave up being a theoretical agriculturist in Africa to become a practical farmer in England.

He was very enterprising, always devising new schemes and technical inventions. He built up a nice little business selling cream from Jersey cows. He designed his own system for pasteurizing the cream, which was full of flashing colored lights and controls. My fascination with computers probably stems from my father's interest in designing cream pasteurizers and other ingenious devices. While I worked on the farm during school holidays, I never took to it terribly well.

Were you an exceptional student?

I was quite unexceptional. When I first went to boarding school in England at the age of eight, I easily reached the top of my class, but then

I dropped behind and became very average, which is probably owing to the fact that people have spurts and lulls in their rates of development.

How did you become interested in biology?

Many of my colleagues came to it through being bird watchers or bug hunters or flower collectors. My father was a flower collector. But I was none of these things. My interest was more philosophical. I was fascinated by questions that traditionally have had religious answers.

Much of your writing is devoted to giving scientific explanations for phenomena that are usually accounted for by the "hand of God."

I am a fairly militant atheist, with a fair degree of active hostility toward religion. I certainly was hostile toward it at school, from the age of about sixteen onwards. I mellowed a bit in my twenties and thirties. But I'm getting more militant again now.

It was a mind-blowing experience to discover Darwinism and realize there were alternative explanations for all the questions with traditional religious answers. I became irritated at the way the religious establishment has a stranglehold over this kind of education. Most people grow up and go through their lives without ever really understanding Darwinism. They spend enormous amounts of time learning church teachings. This annoys me, out of a love of truth. To me, religion is very largely an enemy of truth.

When did you discover Darwinism?

My father told me about Darwinism when I was sixteen. At first I didn't believe it. I needed a bit of time to come to terms with it. I must have encountered Darwin's theory before then, but I don't think I understood it. I believed in evolution, but I somehow thought there was a guiding hand to the course it took. But in Darwinism there is no guiding hand, only natural selection.

Did you have good textbooks in biology?

There was no such thing as a good biology textbook in the 1950s. The textbooks treated biology like a branch of Latin grammar; the more difficult it was to learn the better. But I had a very good zoology teacher, who cared passionately about his subject. He sort of got me into Oxford, because I wasn't expected to get in. I wasn't in the scholarship stream at school, but he gave me extra tuition. I scraped through by the skin of my teeth and then never looked back after that.

What happened when you got to Oxford?

I took to the tutorial system, which at Oxford is a one-to-one relationship between tutor and student. I wrote enormously long essays that I'm sure, if I read now, I would think were dreadful. But they were original. I always took my own line on things, and I really thrived in the Oxford intellectual environment.

When did you meet Niko Tinbergen, the inventor of the modern science of animal behavior?

In my next-to-last term as an undergraduate, I was sent to him for tutorials. He was very complimentary about my essays, and when I came to think about what I might do with my life—not having given it much thought—I decided to stay on as a graduate student. Tinbergen had a very balanced view of ethology. He believed in studying the control and development of behavior, as well as an animal's survival behavior in the wild. He had me study the old question: To what extent can behavior be called innate, as opposed to acquired? I researched the development of pecking in chicks, setting up a simple mathematical model of choice. What was going on inside the bird's head when it made a decision to peck? It didn't solve any important problems, but it was methodologically interesting. People thought it was fairly clever. I was noticed in the animal behavior field and was fairly in demand for jobs.

What do you remember most about Tinbergen?

On a personal level, his great kindness, the fact that he was always smiling; and professionally, his dogged determination to clarify. He would never tolerate fluffy or unclear thought. He was a very good influence on groups working together, because he insisted on clarity, sometimes to the point of exasperating people. This persistence about clarity was something I valued and tried to practice myself.

What is ethology?

It's the scientific study of animal behavior. You have to add that it comes out of zoology rather than psychology, because the animal is studied for its own sake, not for what it can tell you about human behavior.

What did you do after working with Tinbergen?

I left Oxford in 1967 for a teaching job at the University of California at Berkeley. I didn't apply for the job, and I hadn't even published anything, but I had a flattering offer on the strength of my thesis. My first impression on landing at the San Francisco airport was seeing hundreds of young men in uniform. One was abruptly reminded of the fact that this was a country at war. California impressed me with its six-lane highways, cars overtaking each other on both sides of the road, and a general feeling of bustle and busyness. I was fairly overawed really, and it was an influential time of my life.

Did you discover any memes in California that you wouldn't have found by staying in England?

I had just married my first wife, so that time for me is bound up with her and the early days of our marriage. We got involved in Berkeley politics and the Vietnam antiwar movement. But I can't put my finger on anything I took away from Berkeley and could say, "That was a formative influence on my life." I suspect there aren't very many of these on anybody's life.

Why did you go back to Oxford?

Tinbergen came out to visit us and said he was very keen to revivify the Oxford ethology group during his last years before retirement. He was trying to gather together a really strong team to send him out with a bang. I was flattered to be asked to join that, but I didn't come back to a permanent job. It was a research position, with only the possibility that there would be a job later on. As it turned out, a job became available almost immediately, which is the one I now hold.

You divorced and remarried on returning to Oxford?

Yes. I'm not one of life's success stories in that department, I'm afraid. I've mentioned my first wife, and I shouldn't just let that pass. She's a very good scientist herself, Marian Stamp Dawkins, and she's had an enormous influence on my life. We did everything together, and all our work in those days was either collaborative or done in consultation with one another. We published two or three papers together, but that doesn't indicate the extent to which we worked together and thought together. She's still in Oxford, and we are still very good friends and colleagues.

I notice in your writing that you avoid the word "sociobiology." Are you a sociobiologist?

I suppose I am, but when I wrote *The Selfish Gene* I hadn't even heard of sociobiology. It's a coincidence that E.O. Wilson's book *Sociobiology* and mine came out at roughly the same time. Because of the political backlash against Wilson's human speculations, the word "sociobiology" came to have a certain notoriety, and when *The Selfish Gene* arrived in America a bit later, some people mistakenly saw it as part of a movement Wilson had started. I resented being thought of as jumping on a bandwagon that I hadn't even heard of when I wrote the book. I also saw no need to coin a new word. I felt I was an ethologist and that Wilson's book, like mine, was about ethology. However, we seem to be stuck with the new word. So I don't mind being called a sociobiologist, unless people assume I believe things that I don't. Sociobiology has become a red flag word for political activists who think it's about the genetic determination of human behavior. I'm interested in natural selection, and genes are important only because natural selection can't work without them.

Because you talk about animals being "programmed" in the same way that computers are programmed, aren't you sometimes mistaken for being a genetic determinist?

People make two mistakes when they hear you talking about a gene "for" tying shoe laces or some other kind of behavior. They think it means that X is inevitably and irrevocably determined once you've got the gene for it, and this gene is the only one that influences X. Neither of these is true. I use the analogy of baking a cake. You can liken genes to the recipe. You follow the recipe in the book, and then what comes out of the oven is a cake, but you would never break the cake up into bits and say this fragment corresponds to the first word in the recipe. There's no such thing as a gene "for" a bit of cake. On the other hand, if you change one word in the recipe, what comes out of the oven is a different kind of cake—sweeter, moister, or whatever.

Why do people get upset when they hear you talk about a gene "for" tying shoelaces?

Throughout history people have misused genetics by saying we have genes "for" criminality, "for" aggression, "for" homosexuality. If you think this implies an irrevocable commitment to criminality, the political implications would be profound. You could put people in prison because they had certain genes. But this has nothing whatever to do with the reason why I talk about genes. Natural selection would be nothing without heredity. If peas with a gene for tallness are better at surviving than peas with a gene for shortness, then natural selection can choose tallness in peas. Natural selection works fundamentally at this level of individual difference.

Did your gene-oriented view of the world get attacked as soon as* The Selfish Gene *appeared?

There is a widespread view that *The Selfish Gene* met with a lot of flak when it was published. But it got the sociobiology backlash only after a hundred favorable reviews. Ironically, the book is seen as an extreme version of sociobiology, because its thesis about the gene as the unit of selection is phrased in far more radical language. Sociobiology is hardly radical at all in the way it talks about natural selection. It's almost group selectionist. But there is no genetic determinism in *The Selfish Gene.* Quite the contrary. As you yourself pointed out, it ends up talking about rebellion. It doesn't matter if you've got genes for tying shoelaces. They are not little dictators.

The Extended Phenotype *strikes me as your most provocative book.*

It's the most radical and by far the favorite of my books. It's the one where I feel I made the most original contribution. But because it was written for an advanced audience, it isn't as widely read as the other two.

***The title is a bit difficult. Maybe you should have called it* The Gene Ascendent *or* Evolution by Remote Control.**

If I ever wrote a popular version, I was thinking of calling it *The Long Reach of the Gene.* The book is about genes manipulating the world outside the bodies in which they sit. It sounds as though I'm talking about new discoveries, but these are familiar phenomena, things like parasites changing the behavior of their hosts and beavers building dams. I found a different way of seeing familiar facts, which startles people—it even startled me—into making new discoveries. If you are a field worker going out to the Serengeti to study lions, and you've read *The Extended Phenotype* and really taken it on board, you don't ask how the lion is benefiting its genes by certain behavior, you ask, "Whose genes is the lion benefiting?"

How intelligent is the gene?

There is no cognition in the gene. People who take gene selection over the top will say things like, "What were the genes doing, letting us develop contraception? It's not doing them any good." But genes don't have foresight. *The Extended Phenotype* advances the thesis that genes have greater power than they're normally credited with. But that's in the sense of reaching outside the body and manipulating other individuals, even the world at large. That's not a new theory; it's just an upside-down way of expressing what's already known.

Take a beaver dam. It's presumably an adaptation for the good of the beaver. So genes for dam building are favored in beaver gene pools. But the phenotypical [external] character the gene is influencing in order to benefit itself is actually the lake. So I regard a beaver lake as the phenotypic expression of beaver genes by the same logic we regard the beaver's tail as a phenotypic expression of the beaver. By extending that logic, genes in one body can be said to influence other bodies.

One example of genes manipulating other bodies to ensure their survival involves parasites. You even say that genes themselves can be thought of as parasites.

If the body is a temporary throwaway survival machine, with the genes constantly changing partners, why does it function as a coherent whole? Why do all our muscles and sense organs work together cooperatively? The answer is that our genes have only one means of going on down the river that runs from generation to generation. Since they can only leave via a sperm or an egg, they have to cooperate with one another to get into one of those vessels. But if some of them could find another way of getting into the future. . . .

Suppose genes in cells lining the nose caused you to sneeze them into the air, where they were breathed in by somebody else and inserted into the DNA of that person's nose. This would be a very satisfactory way of getting down through the gene generations. These sneezing genes I'm talking about could well exist. But then, by definition, we no longer call them our own genes. We give them a separate name and call them viruses. In more orthodox kinds of parasites, like worms in snails, we tend to view host and parasite as two different individuals. I prefer to say the only reason worm and snail tissue pull in different directions is that they have different exit routes from the body they share. If they had the same exit route—if the worm put its genes into the sperms or eggs of the snail—then they would have exactly the same interest in the future.

Are there examples of parasites hijacking the reproductive mechanism of their hosts?

Some bacteria pass on their genes to the next generation in the eggs of insects. Since they have exactly the same interest at heart as the host genes, I predict that someday they will cease to be parasites. The very concept of "parasite" genes and "own" genes will no longer have any meaning. Maybe we should look upon all genes as mutually parasitic. Local clubs of them cooperate only because they have discovered the same joint solution to the problem of getting into the future.

Communication is another tool used by genes to manipulate bodies. Can you explain how this works?

In the conventional view, animal communication is thought to be informative. But I think it's actually analogous to advertising or propaganda. When I look for human parallels to bird song, I think about speeches by Hitler or Billy Graham, or advertisements for cigarettes, which use subliminal salesmanship to persuade you against your better judgement to smoke.

The other half of the equation is mind reading. Animals mind-read each other when they look for telltale behavioral signs of what other animals are doing. If you see somebody surreptitiously looking at his or her watch, you can tell something about that person's internal state of mind. Animals do that kind of thing all the time. Manipulation is one response to having your mind read. If you deliberately and ostentatiously look at your watch, you can manipulate the behavior of the person watching you. It's a kind of arms race on both sides.

What do you mean by arms race?

It's a race in evolutionary time between predator and prey, parasite and host. A rabbit runs a race with a fox, but the lineage of rabbits also runs an arms race with the lineage of foxes. They are constantly improving their armament: running speed, muscles, and sense organs.

Is there an arms race between men and women? You imply this is inevitable given the fact that the female sex cell (egg) is larger than the male sex cell (sperm).

The gamete-size argument is an aspect of *The Selfish Gene* that worries me a little bit. I'm not sure if it's wrong, or if there is a better way of putting it. As long as you've got two sexes, call them A and B, there is an inherent instability that tends to lead one of them to become malelike and the other femalelike. There's got to be some variation in reproductive success, which will have a self-reinforcing effect. The sex with small gametes will move towards becoming highly competitive and fighting other members of the same sex, while the one with large gametes goes to the opposite extreme, spending time and energy on nurturing young. But the difference in gamete size is the *consequence* of this fundamental asymmetry, rather than being the original motor.

So what is the cause of sexual inequality?

Sexual inequality results from this fundamental difference in reproductive success. Take an elephant seal. The most successful male elephant seal is hundreds of times more successful than the least successful male elephant seal. But this is not true of the females, who are nearly always closer to the average. This means there's more at stake for a male in being competitive. I'm suggesting that inequality is self-reinforcing. If you start out with a slightly greater gain from fighting than nurturing, selection favors an even stronger tendency to fight. So you get a runaway split, with the "fighters" becoming male and the "nurturers" becoming female. The nurturers don't bother to fight, while the fighters spend so much time competing that they have none left over for rearing young.

How do you characterize these differences between male and female behavior?

Males are high-stakes, high-risk gamblers. The bull elephant seal who gets a harem fathers an extremely high percentage of the next generation. If he loses, he fathers none. The female is a low-stakes, low-risk gambler, hardly a gambler at all. She takes the safe, middle-of-the-road strategy.

How has the work of Sarah Hrdy and other scientists studying female choice altered this view of females as passive, middle-of-the-road players?

I would never call females passive. The whole Darwinian theory of sexual selection is based on this idea of female choice. Male pheasants and peacocks compete for how much they can appeal to females. But it's still the same idea even if females are choosing. The successful males are the ones that get a lot of females. In this case, they get them by appealing to females, via female choice.

You say in* The Selfish Gene *that it's in the male's interest to inseminate as many females as possible, while the female wants to secure a stable mate.

The male who behaves like a sultan can have many more offspring than any female possibly could, simply because males don't get pregnant. So there's no virtue in females having a highly developed sense of peripheral vision, unless they use it in some subtle way, like sneaking copulations and confusing the issue of paternity.

From the perspective of the gene, how can you explain why there are two sexes to begin with?

Suppose that a human female mutated to become asexual. This asexual female would be capable of producing as many offspring by herself as she could with the aid of a male. But each one of these offspring would contain all of her genes, instead of only half of them. Take a species where the male doesn't do any work raising the offspring, like a pheasant. The clutch of eggs that a female can rear is exactly the same whether she has a male to help her or not. An asexual female will rear twice as many of her own genes.

Asexual reproduction, on the face of it, has a twofold advantage over sexual reproduction. So when we ask the question, "What is sex good for?" we have to think of a colossal advantage large enough to outweigh this twofold cost. It's an unsolved problem. It's being tackled by mathematical models of various kinds, and it's a very active and exciting field of theoretical research.

Why do you think sex evolved?

People reading this may be a bit puzzled, because they probably learned at school that sex increases variety. The problem with this is that, unless you're careful, it's a species-level advantage, which is the wrong way of thinking about the problem. Species practicing sexual reproduction have a greater capacity to avoid extinction. They will have a larger range of types to call upon when the going gets tough. An asexual species is likely to be entirely uniform. It may be doing very well in the present climate, but as soon as there is an ice age or a major drought, the whole lot get wiped out.

What is the advantage of sex at the gene level—the level you say counts most?

People have tried to think of ways in which an individual can be more successful at reproducing, through variety. This argument seems to depend upon a world in which there is competition between siblings. If our offspring compete among themselves, and the winner of my little competition takes on the winner of your competition, then you can see the advantage of having variety in your family. You want to have a champion who has been selected from among a great variety of contenders, and sex does provide that.

It also becomes easier to understand why sex exists if the environment is changing so rapidly and in such perverse ways that it becomes impossible for a species to evolve to perfection. If the environment is running harder than the species—which happens when it consists of predators and prey and parasites who are themselves evolving so that the species can never settle down into a cozy nirvana of perfection—then sex can have an advantage, a substantial one.

How do these truths about animal behavior affect humans?

I would be aghast at any suggestion that these sex differences should be regarded as prescriptive for human behavior. I don't find them a very helpful key to understanding behavior of the sexes in my own social life. Humans are really rather good at rising above these things.

The Selfish Gene *strikes me as a young man's book. You were willing to generalize about behavior in a way your subsequent books haven't.*

I was a little naive about the possibilities for misunderstanding. I was also naive about the tendency of people to be mainly interested in humans, when I wasn't. I suppose my other two books were written in light of experience gained from the first.

You and Stephen Jay Gould recently debated the theory of evolution before an audience of a thousand people in Oxford. What was the nature of the debate?

I advocate the gene as the level at which natural selection acts, while he advocates a variety of higher levels. Gould wants to be catholic in his approach, while I want to be rigorous. Natural selection has to work on something that's self-replicating, and your individual organism is *not* a unit of selection. The debate was cordial. It was hard-hitting. But we both went away feeling just the way we did when we came in.

In* The Extended Phenotype *you claim to have led a scientific revolution in the way we think about evolution, while on the other hand, you look like a defender of neoorthodox Darwinian theory. Is your work revolutionary or conservative?

I'm orthodox in the sense of being neo-Darwinian, but I see the implications of what neo-Darwinism has to mean in a different way from most people. I am fundamentally orthodox in the matter, but radically unorthodox in the method of presentation. And I like to hope that this change in method—in the analogies, metaphors, and images used to describe things—moves over from being a trivial matter of how you express something to being creative in its own right. It is this hard-to-tease-apart mixture of orthodoxy and unorthodoxy that perplexes you.

Gould is also fundamentally orthodox, but he expresses his ideas in an unorthodox way. The difference is that he chooses to *describe* himself as unorthodox. He uses phrases like, "The modern Darwinian synthesis is effectively dead." I don't think it is, and I see myself as squarely in the neo-Darwinian tradition, even if I reexpress it in unfamiliar imagery.

Darwinism is usually defended by recourse to the fossil evidence. But you defend Darwinism at the level of complex systems—in other words, by means of logic.

We've almost got beyond the stage of needing evidence. Darwinism is the only theory that works. Fossils are interesting for discovering the actual historical course of evolution and how it happened. But they're not a very good source of evidence for proving that Darwinism is true. They prove that *evolution* is true, but no sane person doubts that evolution is true.

In* The Blind Watchmaker *you drop the gene's-eye view of the world to give a more orthodox interpretation of evolution.

I've switched topics. Where my earlier books dealt with social behavior, *The Blind Watchmaker* is about Darwinian design by natural selection. Design is as much a matter of engineering as a matter of behavior. The book deals with the question of how we recognize design, what it looks like anywhere in the universe and by what processes it can come about. This is a more general subject, because behavior is a specific example of design.

Why have you stopped writing about social behavior?

I didn't feel I had another book to write along the lines of *The Extended Phenotype*. I had given a lot to that, and I had nothing more to say, not just yet anyway.

You spend your days sitting in front of a computer, wandering through biomorph space. What is this?

I've designed a program called *The Blind Watchmaker,* which gives a very realistic model of evolution. It shows the power of cumulative selection to generate an almost endless variety of quasi-biological forms. The fundamental form of all the biomorphs is a branching tree. But by breeding through generations, they can assume quite amazing shapes. The program is the equivalent of embryology. It allows you to change genes systematically, to collapse and stretch limbs in ways that resemble genetic influences on embryology.

Why did you call these shapes biomorphs?

I borrowed the term from my friend Desmond Morris, who uses it to describe the shapes in his surrealist paintings. He claims biomorphs evolve from canvas to canvas and so take on a life of their own. Now that I've got into color and more solid-looking forms, mine are starting to look more like his. I keep altering the biomorph program so that it embodies the biological changes I'm thinking about—changes in segmentation, symmetry, and so on. I keep switching back and forth between real life and the computer.

How much time do you spend programming?

I had a great spurt earlier this year when I developed the color version and added lots of new genes. It tends to be solid for a couple of months, and then I'll drop it completely. I often get up early and work a lot before breakfast. The middle of the day is largely taken up with public things, teaching, tutoring, and so on. Then I start working again in the evening. I don't have a television; so that's one major distraction removed.

How did you learn about computers?

It's always been a bit of a vice. Frankly, I've wasted a lot of time. I had an Apple II computer in the early days, when there was no good word processor. So I wrote my own in machine code. It was an enormous undertaking, and I'm probably the only person in the world who ever used it.

What are you working on at the moment?

I'm thinking a lot about embryology. What may have happened is that embryologies have evolved that are good at evolving. Certain kinds of embryology are good not only for individual survival, but also for giving rise to further evolution.

I was led to thinking of this through playing with the biomorphs, and recognizing that each version of the program has its limitations. You have to go back to embryology and change the fundamental form before you can get a whole new flowering of evolution. That's probably what happened in the history of life. Every now and then there's been a major change in the way the embryology works, which maybe didn't have any immediate benefits as far as survival was concerned, but which opened up floodgates of further evolution. So today we have embryology that is very good not only at surviving but also at evolving.

Farouk El-Baz

✦ ✦ ✦

WHEN Farouk El-Baz drilled into a recently discovered tomb under the Great Pyramid of Egypt, he was not searching for archaeological remains, ancient treasure, or a Pharaoh's missing mummy. He was looking for air. "This would be a really fabulous discovery," says the founder of the Center for Remote Sensing at Boston University. "Think how much you could learn about global warming and other changes in the atmosphere from studying a 5,000-year-old sample of Egyptian air."

Instead of air, however, his remote sensors and camera's eye found the "Pharaoh's curse"—a sign from the ancient Egyptian kings that they are still jealously guarding their secrets. But he also found a cedarwood boat that could be the Pharaoh's funeral bark, or possibly the final resting place for his mummy. The boat is companion to another craft discovered under the Great Pyramid in 1954. Reconstructed from more than a thousand disassembled pieces of wood, this previously discovered boat is the oldest in the world, and its design reveals a level of technical skill as great as the pyramids themselves.

Soon after it was excavated from what was thought to have been a hermetically sealed chamber, the Pharaoh's boat began to deteriorate. That's when the Egyptian Antiquities Organization (EAO) contacted El-Baz. As an expert in remote sensing technologies, which he developed as scientist in charge of the Apollo moon missions, could El-Baz figure out what was happening to the boat? He decided to drill into the second unexplored chamber under the pyramid. Should it, too, be hermetically sealed, the chamber would reveal the best atmospheric conditions for storing the reconstructed boat. It might also contain a boat of its own, and that precious sample of ancient air.

Accustomed to exploring upward into space, El-Baz, for his archaeological work in Egypt, needed to delve downward into history. "Either way you go," he says, "it's a fantastic journey." He called on friends from the Apollo missions and other experts in imaging systems to design the equipment and methodology for a new kind of "nondestructive" archaeology. Whether studying the Sphinx, pyramids, tombs, or other sites in his native country, El-Baz disturbs the objects as little as possible, and he leaves them in place. "The era of archaeology as high-class grave robbing is over," he says. "It's time we realize that archaeological sites themselves are often as important as the objects they hold."

One of a family of nine children, El-Baz saw his first pyramid when he was twelve. There were no restrictions in those days, he recalls, "so we raced each other to the top and went back many times, climbing all over the immense structures and crawling inside." After getting a master's degree at Ain Shams University in Cairo in 1959, he received a Ford Foundation scholarship to study in the United States. Trained as a geologist at the Missouri School of Mines, MIT, and the Max Planck Institute for Nuclear Physics in Heidelberg, he is today one of the world's—actually the solar system's—great explorers. He led twelve expeditions into the Sahara, as well as other African, Asian, and American deserts, to study their formation, before focusing on another dry climate: the surface of the moon.

In a classic test of remote sensing technology, El-Baz figured out where the first manned mission to the moon should land, what the astronauts would be walking on when they stepped out of their space capsule, what samples they should bring back, and how they would collect them without contamination. El-Baz became geology instructor-in-residence for the astronauts. Judging from the testimonials lining his office walls, he was a great teacher. "*Marhaba ahle el-ard, min* Endeavor *elaykum salaam,*" said astronaut Al Worden on finishing his twelfth orbit around the moon. Ground control panicked, thinking his space capsule had been hijacked by Martians. But Worden was speaking Egyptian for the benefit of King Farouk, as El-Baz was affectionately known. "Hello, people of Earth, greetings from *Endeavor.*"

El-Baz has analyzed photos of the windswept surface of Mars, which "looks like the desert of my childhood," says the now-naturalized American citizen. From satellite photos of the Sahara, he has discovered vast reserves of groundwater and soil ripe for reclamation. In 1975 he directed Earth observations on the Soviet–American Apollo–Soyuz spaceflight. And before Anwar Sadat's assassination in 1981, El-Baz was science advisor to the Egyptian president. "Farouk is always bringing his expertise back to his native region and applying it to important problems," says a colleague. "He's looked at as the ideal Egyptian scientist."

Talking with El-Baz in his Boston University office, I faced a wall-sized mural of the space shuttle Columbia flying over the Red Sea. The remaining walls were filled with Landsat images of Boston and Cairo, photos of the world's deserts, and other memorabilia from the Apollo program and Smithsonian Institution, where El-Baz served for ten years as the founding director of the Center for Earth and Planetary Studies at the National Air and Space Museum. Beneath his tortoiseshell spectacles, El Baz's sunburned face often broke into a grin as he recounted adventures from a lifetime spent exploring this, and other, worlds. Six hours later, with many good stories left to tell, he said, "You'll have to come back again for another visit."

✦ ✦ ✦

Why were the pyramids built?

I think building pyramids was equivalent to Roosevelt's WPA projects or the Apollo space program. The pyramids would distinguish Egypt from everyone else, unite the country, and keep the masses working on a fantastic project. The were built over a short period of time in an event not to be repeated. But each individual pyramid took at least twenty years to finish, indicating it was a part-time job. Otherwise the ancient Egyptians would have finished the work quicker.

For three months of the year when the Nile Valley floods, you can't do any farming. So the Pharaoh hands out grain to people who come work for him. He puts together a melting pot of northerners and southerners working on a project they can feel proud of. They intermarry and unite to build a country. Egypt became a country then—and the boundaries have not changed for five thousand years.

So the pyramids weren't built by slave labor?

This Hollywood version of slaves laboring under the whip simply doesn't work. The pyramids would not be as elegant as they are. To finish their quotas, the workers would have filled in areas with junk rather than perfect squares. They wouldn't have been proud of their workmanship, as you see on the many blocks signed with the names of crews that made them: the "perfect cutters," the "falcons," and so on. These were teams competing against each other, looking for ways to do the job better. There was singing and joy and pride in building these beautiful structures. We found similar markings when our TV camera began transmitting out of the boat chamber.

But doesn't the Hollywood version date back to Herodotus, the oldest non-Egyptian source on pyramids, who describes the Pharaoh Khufu as a tyrant who forced his subjects to work as slaves?

Herodotus was a tourist who listened to anybody who'd tell him a story. And Egyptians from ancient times have been storytellers, the juicier the better. Along comes this guy who knows nothing about the history of Egypt. He can't speak the language or read the writing on the walls. He sits down with his camel drivers and some fellow Greeks, who have their own motives for making Khufu look like a tyrant. Remember, the Greeks were pushing to take over Egypt, which they eventually did. Since the poor Egyptians were tyrannized by their rulers, says Herodotus, maybe the Greeks can come in and fix things up, give them some philosophy, civilize and liberate them.

Where did the Egyptians get the idea to build pyramids?

The shape is based on what they saw in the open desert. I was flabbergasted when I began exploring places that humans may not have seen for five thousand years. Out in the middle of the Great Sand Sea you find conical structures where nothing should be standing. The wind is so fierce that everything around them has been totally leveled. Did the ancients notice these forms and realize they were durable?

A great architect, Imhoteb, invented the idea of building repeated mastabas, or steps, that get smaller the higher you go. He built the first pyramid for the Pharaoh Djoser at Saqqara around 2630 B.C. Architects after Imhoteb realized they could build pyramids without steps. They covered the blocks with mortar providing a layer of white plaster to color. A pyramid with a smooth surface was a fantastic billboard. Excellent bureaucrats, the ancient Egyptians left nothing unwritten. They covered the surfaces of temples, obelisks, tombs, and clay tablets with their history. Because the history of the pyramids and Pharaohs who built them was written on the outside, these overwhelmingly massive objects could be read from miles away. When the Pharaoh's mummy was sailed across the Nile to be buried, with music playing and these awesome objects shining in the desert, that's when you cried and gave way to religious feelings.

Where are the mummies?

In the middle of the Great Pyramid of Khufu, in what's believed to be the king's burial chamber, lies a fabulous granite sarcophagus. But the king's body has never been found. We have no mummies from the Old

Kingdom: guys 4,600 years old. Actually they might be older by another four hundred years. Colleagues are coming up with new dates that push all Egyptian history back four hundred years. Most of the mummies we have are 3,500 years old or younger. Mummification took place in stages, and only the Pharaoh got the fullest treatment. Nobles got lesser treatment, and commoners little or none at all. We have plenty of skeletons and bones from 5,000-year-old Egyptians, but no perfect mummies.

I have a wild idea.

I like these.

What if Khufu's mummy is lying in the boat you discovered?

Oh, absolutely. If not in this boat, then somewhere in the boat pit. Khufu and the Pharaohs after him may have built these magnificent pyramids to draw attention away from their real tombs, where their bodies would lie undisturbed. The burial chamber for the guy following Khufu, the builder of the second pyramid at Giza, has never been found.

The American physicist Luis Alvarez thought he could find the chamber with cosmic rays. He entered the only miserable compartment we've found, at the base of the pyramid, and pointed his ray detector at the sky. Rays coming through a vacuum should travel at a different rate than those moving through stone. That's how he hoped to locate the chamber. But he got more noise than signal. His equipment wasn't good enough.

Maybe not in my lifetime, but someday somebody will come along and do the job right. The technology is developing so fast that there's no need to rush into one of these discoveries. We have to leave some things for the next generation, especially things that right now look a little too difficult. Don't fuss with it until somebody develops an instrument, a methodology, that will allow us to make this discovery in a nondestructive way.

What happened to the missing mummies?

All the mummies we have today were found in the twentieth century, most of them saved by Egyptian priests. Before the country was invaded by the Hyksos, the priests took the mummies out of the pyramids and stored them in a cave in the Eastern Desert. When the cave was discovered early in this century, we knew from the markings on the coffins that we'd discovered the royal mummies, including Ramses II, the greatest Pharaoh of them all. The single major exception is Tutankhamen. He was found interred in his tomb, and after President Sadat said it was a desecration to remove his body, he was put back in his coffin.

Because the pyramids are geologically aligned, some people claim they functioned as observatories.

The pyramids are so perfectly oriented, north, south, east, and west, that scientists have used them to measure the rotation of the African continent over the past five thousand years. But everything the Egyptians built is perfectly oriented: the Luxor and Karnak temples, Abu Simbel, even the little tombs in the cemeteries.

Why are the boats that have been discovered in the pyramids called "solar" boats?

Tomb paintings depict the ancient gods holding boats in their hands and raising them toward the sun disc. Is this a symbolic boat for transporting the soul of the Pharaoh across the sky? Two boats would be required: one for the journey by day and one by night. Some think the boats actually transported the Pharaoh's mummy. There is no proof these boats were used in his funeral, and there is no gold on them or other decorations, but they look like funerary boats. Many boat pictures appear in tomb paintings, and there are also lots of boat models in tombs—like the two small boats found next to Tutankhamen. The boats are obviously connected to death or the journey after death, including a possible return to life.

There was no question in the ancients' minds that they were going to come back to life, not reincarnated, but actually living again in the same body. That's why whenever someone of notoriety was buried, he or she would be surrounded by milk, grain, bread, and vegetables. The higher in rank, the more amenities. And of course, you needed a boat. The ancient Egyptians lived on the east bank of the Nile, but buried their dead on the west bank, with the setting sun. They transported the bodies by boat, so when you came back to life, you'd have to cross the river again to be among the living.

What makes you think the solar boat carried the Pharaoh's mummy?

The solar boat, the one that's been reconstructed, is the oldest ever found. It's long and sleek, and the wood is curved into a magnificent melonlike shape. At one hundred forty-two feet, it's forty-two feet longer than the Viking ships that crossed the Atlantic. The main cabin has two rooms, maybe one for the mummy and the other for people accompanying him on his journey to the west. I climbed inside to study its construction and saw how the planks were tied together with ropes threaded through little holes. Nails weren't discovered for another fifteen hundred years. No iron or steel holds the boat together. The engineers who designed it rank among the best builders ever. I saw where the ropes left marks on the wood, which means the boat once floated.

What other theories explain the existence of the solar boats?

They could be funerary vessels or symbolic boats for visiting the sun god or boats required for life after rebirth. But nobody knows why we have two boats. Five pits surround the Great Pyramid, but three of them, also in the shape of boats, were plundered in prehistory. One of these contained a tiny sliver of wood painted gold, so there might very well have been five boats instead of two.

How were the solar boats discovered?

In 1952 King Farouk took a Saudi prince out for a drive around the pyramids, but they got stopped by a pile of rubble on the south side. The King called the head of the antiquities department and told him to clean up the mess. It took him two years—he was being careful not to disturb any cultural remains—and when he reached the bottom of the pile, he noticed some straight-edged rocks that looked like they had been fashioned by human hands.

"There is something hidden here!" said the archaeologists. "Maybe the mummy of Khufu!" They removed the first block of limestone to open a chamber that may have been hermetically sealed. Out of it came the fresh smell of cedarwood imported from Lebanon five thousand years ago, and in the pit they found a wooden boat, disassembled into one thousand two hundred twenty-four pieces, but perfectly preserved. It took them thirteen years to put the boat back together again and another five to build a museum for it. Then they noticed the boat was deteriorating. That's when they came to me with the problem of figuring out the best environment for preserving it.

Why did you want to know if the pit had been hermetically sealed?

A pristine sample of air from ancient Egypt would be fantastically significant. We only started sampling the atmosphere quite recently to discover what we think is global warming and an increase in the amount of carbon dioxide. But what if the amount of carbon dioxide in ancient Egyptian air was *greater* than it is today? This would present quite a puzzle. What if it contained chlorofluorocarbons? Then the theory that chlorofluorocarbons are produced by volcanos would be proved correct, and we wouldn't be doing as badly as we think we are in relation to the ozone layer.

Where else could you look for old air?

Mine shafts and caves are contaminated by gases coming from within the earth. This limits you to looking for something made by human hands that was either intentionally or unintentionally sealed, like ancient tombs, either in Egypt or China or Central America. You could also try looking in one of these big granite sarcophagi, assuming it was sealed with plaster. The Smithsonian Institution has an empty bottle that was corked one hundred seventy-five years ago, but they won't let anyone open it. Our only samples so far have come from air pockets in glaciers, but their small amount of carbon dioxide doesn't allow for good comparisons.

Describe your attempt to capture a sample of air from the second pit.

To me the composition of the air seemed as important as anything we'd find in the pit. Even if my calculations gave me only a one percent chance, it was worth it. But how were we to analyze this air without touching it? We had to locate the boat underground and drill through two meters of limestone without contaminating the air inside the chamber. This meant we had to drill without lubricants or coolants and slowly enough not to raise the internal temperature of the tomb. The problem was very much like those we handled on the Apollo missions. Sight up all the aspects and tackle them one by one, designing backups to your backups all the way. I thought of it as a microcosmic moon mission. Instead of traveling into space, we were traveling down into history. By modifying archaeology with space-age technology, we were exploring a new frontier. The job had to be done right, because there would be no chance to redo it. So we had to be covered from A to Z.

Two-and-a-half years passed before we started the operation. We began by using ground-penetrating radar, which indicated something was sticking up in the middle of the pit. So we drilled one-third of the way along the surface. We hung a drill heavy enough to cut through six feet of limestone from a wooden scaffold tied together with ropes just as in ancient times. We used a maneuverable camera modified with a fiber optic light source. It was originally designed to check for cracks inside nuclear reactors. The size of its head—eight-and-a-half centimeters—defined the size of the hole we drilled.

Our most remarkable piece of technology was the air lock around the drill. It was designed by Black & Decker engineer Bob Moores, who designed the lunar drill for the Apollo astronauts. On the moon we had to drill a soil sample without using any mineral oils or air pressure that would alter its composition. Here it was even trickier, because we had to drill without mixing the air inside the chamber with the air outside. We kept the drill bit under slightly negative pressure, so if anything broke, air would go into the drill rather than come out.

You also published an article in* Science *asking for advice from the public at large. Did you get any good ideas?

All kinds. But maybe the best suggestion came from a scientist in Boston who called to ask how we were planning to reseal the chamber after we were finished drilling into it. I named an epoxy resin, and then she asked, "Why not seal it the way the ancient Egyptians did in the first place? They used gypsum mortar." She was right.

What happened when you finally broke through?

There was no change in pressure, which meant the damn chamber wasn't airtight. Over half of what we'd done in preparation was useless. People from the EAO said, "Now that it's no longer airtight, let's just open up the chamber." But I said, "This isn't a one-time job. We're trying to prove the technology. We don't know what's inside or what shape it's in." So we continued the way we'd planned.

Besides the solar boat, what did your cameras find?

Pharaoh's curse. Somewhere in every archaeological excavation you find the curse of the ancient Egyptians: revenge for disturbing the dead. For us the curse took the form of a dung beetle. A living beetle meant the mortar sealer between the blocks was opened. We already had marks indicating holes large enough for water to seep through, but now we obviously had holes big enough for an insect.

Do you think the first boat pit was hermetically sealed?

No. We found a lot of freon in the second chamber, in proportions as high as outside, and we've calculated that the same thing must have been true for the first pit when it was opened.

What are your plans now for the solar boat?

I've suggested to the EAO that we drill back through the top of the chamber and use our cameras and other devices to figure out what leaks. Then we can reseal it and fill the chamber with an inert atmosphere, something like nitrogen, that will keep bacteria from destroying the wood. Through the hole we've drilled we can install a TV camera—something like the original we used for the exploration—and hook it to a monitor in the museum. Visitors will be able to move the camera back and forth over the boat, which will be lying in place exactly as the ancient Egyptians left it.

Why did you come up with the idea of nondestructive archaeology?

Many archaeologists think I'm nuts. They say you have to feel an object, hold it in your hands. But some of the younger ones understand the idea and support it. All over Egypt you can see land that has been badly disfigured by archaeological excavations. The hills at Luxor are covered with scars from fifty years ago. It's as if someone took a giant rake and scraped it down the mountains. This practice of digging holes every meter and covering the landscape with rubble is a stage that's passed.

Today we have the technology to figure out in advance where to dig, if that's what you want to do. No more fishing expeditions. Secondly, you should know in advance what you're going to find and be prepared to handle it. Who's going to touch the object? How are you going to preserve it? What kind of environment should it be stored in? I believe that in some cases you can make your museum out of the site itself. We already have one boat, and I can tell you right now what the second boat will look like, so why take it out of the ground?

To find out if Khufu's mummy is inside?

We may not have them right them now, but in the future we'll have plenty of ways to "see" if there's a mummy inside the boat. Wood responds differently than bone, so I can imagine many techniques for distinguishing the two. So why do you want to take out the boat?

To learn how it was put together?

We called in all the experts on ancient boat building, including someone who had spent years studying the first boat. From what he knows about triremes, he reconstructed the boat from our pictures. "This plank has a niche in the middle for fitting into this column and because of its length the boat is this big." So I can already tell you the size and shape of the boat.

Most archaeologists would just swashbuckle into the tomb and pull out the boat.

There's no more room for this sort of behavior. You're disturbing the environment and the archaeological site itself, and for what? For the benefit of touching the object? This shows total disrespect for what the object is and a great measure of selfishness. I've always thought archaeologists were high-class grave robbers. The history of Egypt does not speak kindly of all the people who've gone there and plundered the land of its most magnificent artifacts. The beard of the Sphinx and the Rosetta Stone are in the British Museum. The head of Nefertiti is in Germany. The best obelisk is in the Place de la Concorde in Paris.

In the West you think of archaeologists as exciting explorers of faraway places. This would be OK, if the objects they discovered were left in place. But from the point of view of the people hosting these fascinating discoveries, they see their cultural heritage being stolen. What gives you the right to plunder Egypt to fill up your museums in New York and London? Maybe these objects mean more to me than to you. Maybe I'd prefer you left them where you found them.

Archaeologists would say you Egyptians were sitting on treasures you didn't recognize or know how to preserve.

"Hey, Egyptian," the archaeologists could say, "we've found this face of Nefertiti. It's the most amazing thing in the world, but it's going to deteriorate if you don't put it somewhere safe." Rather than assuming the ignorant native won't take care of it, let me figure this out for myself.

How important is this idea of nondestructive archaeology?

It's potentially as important as the technique of age dating. Archaeology only developed as a science when it could tell us the relative sequence of events in human history. Before that, the only truths we knew about the past were those of history and interpretation, and a few hints from the superposition of objects in relation to each other. Then all of a sudden we got this method that uses carbon atoms to figure out the age of a pot shard or piece of charcoal. Since the invention of age dating, we no longer say, "This hearth might be this old because it resembles one over here." We now say, "It is fourteen hundred years old, and this one over here is sixteen hundred and fifty years old."

Archaeology immediately entered the league of the exact sciences. It was no longer an avocation indulged in by rich people with a lot of time on their hands. Because now, if you haven't aged your site, you don't belong in the business. But nondestructive archaeology is an even more scientific approach. We've changed methodology. You don't move into an area and start digging. First you look at it from space photographs or remote sensors nearer the ground. You define the geology and other controls on human habitation. After you've identified the most likely places where people might have lived, you go in with your instruments and bounce rays off the ground. If you find something worth excavating, you limit your digging to a minute area.

It's unfair to label huge tracts of land as archaeological sites and forbid people to use them, which happens a lot in Egypt. I say, "Tell me exactly where these sites are, mark them, and let the people live around them." We're involved in one of these cases right now. We're using remote sensing technology to map an archaeological site on the west bank of the Nile near Nag Hammadi, which once held a pre-Pharaonic community. This was a town existing just before the initiation of statehood. It's an important time span. We have lots of data on people who lived in the Sahara and on the ancient Egyptians, but the link between them, right before the first Pharaoh, is what we need to figure out. This site also lies in an area that should be agriculturally developed, and I told the minister so. While they begin building roads, I'm going to spend part of the year figuring out what locations should be kept off limits.

How does this relate to your training as a geologist?

Geologists are one reason why archaeology took so long to develop into a science. The rocks that geologists like to play with are tens or hundreds of millions of years old. Geologists generally keep their hands off anything less than a million years old, so we've neglected the relation between geological conditions and human habitation. Civilization occurs with geological consent. You have to have arable land, water, and sunlight. These are *geological* conditions, but geologists tend not to think in these terms. So they fail to supply information to ecologists, environmentalists, and archaeologists—people worrying about the story of what happened today and just last night, by geological standards. In the absence of geological knowledge, these scientists tend to make their predictions from a weak base.

Have you studied mummies?

The Boston Museum of Fine Arts has the largest collection of mummies outside of Egypt, and I used to go there a lot when I was a student at MIT. When you stand next to one for a long time and think of it as a person, with a life of its own so many thousands of years ago, you develop tremendous fellow-feeling for this creature. You begin to think, "Isn't it terrible to be exposed like this?" That's why Sadat ruled that the mummies of the Pharaohs on display at the Egyptian museum should be put back in their tombs. I would go even further to say that the tombs themselves shouldn't be disturbed. I say this not out of religious feeling, but because the sites themselves—the way they were planned and executed—have archaeological significance. With remote sensing technology we can get a lot of information out of a site without disturbing it.

Why are there so many mummies in Boston?

Because George Reisner at Harvard was one of the great Egyptologists. Collecting mummies used to be a big thing for tourists to do. You'd bribe an official, buy a mummy, and bring it home to show the kids. After it lay around the house awhile and you didn't know what to do with it, you'd donate it to the Boston Museum of Fine Arts or to Massachusetts General Hospital.

What have you learned from studying mummies?

The two great secrets of ancient Egypt are how they built the pyramids and how they mummified. The second may be even more important than the first. Here's a picture of Ramses II, whose mummy was brought to Boston for an exhibition at the Museum of Science. A lot of remote sensing technology was used to study his body. We know from X-rays, for example, that he had growths between his joints and suffered a great deal from arthritis. This mummy—this man—is 3,200 years old, and we still have his dried flesh, hair, skin, nails, and teeth. Everything is intact on the bones and perfectly preserved. These things are organic. They should have rotted away many moons ago. No matter how you embalm someone today, the flesh disappears after a couple of hundred years, and all you have left is bone.

What technology is best for studying mummies?

X-rays provide an image of bones as well as background flesh. But now we have nuclear magnetic resonance and an array of other diagnostic technologies that are developing very rapidly. In the old days, you'd cut up the mummy and take out chunks to analyze. Now we're pushing to do micro, nondestructive analyses. To study the chemistry of mummification, for example, we use spectrometers for bouncing light off the mummy's surface. You no longer have to snip off a piece to do your chemical analysis.

What do we know about mummification?

The brain and other internal organs were sucked out through the nose, cleaned, dried, and stored in Canopic jars buried next to the body, in case they were required in the afterlife. We know from analyses of Ramses's skin that the body was treated with a mixture of salt and other substances. But we really don't have the slightest idea how mummification actually took place. All we know is that it took a long time. Lesser nobles may have been finished in a week, but someone of Ramses's caliber would have required at least seventy days to complete the process, and today we have a mummy in front of us so lifelike that you can reconstruct the expression on his face the moment he died.

You've also used remote sensing to study the tombs of Ramses's sons and Queen Nefertiti, his wife.

Many tombs in the Valley of the Kings were opened early in this century. Some were kept open, others neglected, and the tomb of Ramses's sons disappeared completely under a pile of rubble. The EAO thought my new black boxes might find the tomb. So I shipped ground-penetrating radar, seismometers, conductivity meters, and electromagnetic sounders to Egypt. We have a slew of instruments for examining the subsurface, and sure enough, we found the missing tomb with a magnetometer, which measures disturbances in the local magnetic field that result from the presence, or absence, of different kinds of rocks. Everything around a void has a magnetic orientation, but a void has none at all. The tomb we discovered held twelve of Ramses's sons, all of whom he outlived in his sixty-seven years as Pharaoh.

I also introduced space-age technology to study the paintings on the walls of Nefertiti's tomb at Luxor. These are magnificent color murals made with minerals and other natural products. They depict Nefertiti wearing a see-through white gown. She's being prepared to meet the dieties who are seated around her wearing the masks of Isis, Hathor, and so on. It's a beautiful setting and these paintings are outstanding, which is why they're worth saving.

We used instruments to detect salts and pockets of moisture behind the paint layer. The lab work is ongoing, but it looks as if most of the damage came from a one-time event, a super-heavy rainfall that flooded the tomb before its discovery in 1904. Water has percolated through the tomb joints, moving salts to recrystallize behind the plaster layer. Because it's not a continuous process, the tomb can be sealed to prevent further degeneration.

Among your discoveries on the Giza plateau, you seem to have solved the riddle of the Sphinx.

I think the Sphinx is a yardang, a wind-sculptured rock sitting on the plateau. After spending years looking at these Sphinxlike shapes in the Chinese and Indian deserts, the Sahara, and other windy environments, I came to believe the Sphinx itself was an actual landform, and rather than removing it when constructing the pyramids, the ancient Egyptians dressed it up with the face of Khafre (Chephren), the Pharaoh who built the second pyramid. This is the only possible explanation for its location, which otherwise is too far down the plateau, with no distinctive relation to the pyramids.

During a trip to the Farafra Oasis in 1977, I found an honest-to-God sphinx standing there in the desert all by itself. I took a picture and said, "That's the proof I need. Anybody questioning this one can see in the background all sorts of sphinxes-in-the-making." When I published the theory, I was amazed I didn't get a lot of hysterical letters from archaeologists. In fact, almost everyone who wrote said they'd seen similar things and felt the same way.

What exactly is the Sphinx?

The way it eroded clearly indicates that the Sphinx is not built of blocks but carved from the rock in place, mainly by the wind. Any strategy for conserving it must answer two questions: What part does groundwater play in its deterioration? What part wind? We need to make a hydrologic model to see how much water via capillary action gets from the ground to the Sphinx's head. We also have to place instruments around the Sphinx to analyze how the wind vortices work. Then we can put the model in a wind tunnel and compute the erosion rates. Before you pump out the ground water, cover the Sphinx, spray it, build wind baffles, or whatever, you have to figure out nondestructively what's happening now and what will happen after the remedy.

Your ideas about deserts are equally controversial. Why do you call the theory of desertification "Bedouin bashing"?

Changes we don't understand are taking place in the earth's dry lands. The term *desertification* gives people the false impression we know what we're talking about. It's easiest to blame the "ignorant native," while in reality these people know more about the desert than we do, and if we want to understand it, we better ask them what they know. The most dangerous thing we can do is settle the nomads because nomadism is the best possible way of living in desert environments. The recent problems are due to the deep wells and settlement schemes that governments and aid agencies have adopted to control the population. These agencies are trying to find people to help, so they must get them some place to help them: "When I come next month with my donated grain, I want to find you *here.*"

What makes deserts expand?

Deserts don't come from goats nibbling at plants. Major climatic changes caused by variations in the amount of energy the earth receives from the sun have been making deserts expand and shrink throughout geologic times. Deserts go back in the geological record two hundred thirty million years, long before human beings ever walked on earth. Only twenty thousand years ago the edge of the Sahara, which today lies near Timbuktu, extended four hundred kilometers southward into tropical Africa. Enormous dunes, now covered with vegetation, are still visible from space. From this perspective, we're living in wetter times and the desert is smaller than before. The most significant piece of the puzzle is the environmental balance at the desert's edge, where life is so fragile that by doing very little you can tip it either way. But each desert has to be considered separately; to generalize is to condemn.

When did you first see the pyramids?

In seventh grade, when my family moved to Cairo. What struck me was the enormity of these things. They're like five-story buildings, as high as the Washington Monument. But it's not just their height that's impressive. It's their bulk. They are absolutely immense structures.

When did you decide to become a scientist?

I was born on January 1, 1938, into a family of seven boys and three girls in the desert town of Zagazig. It looks like "zigzag" misspelled. This was an ancient town that became a major center for the Arabs when they invaded Egypt in the seventh century. My father taught in the *kuttab,* which literally means the "book readers." Moslem kids went there to study the Koran. The best ones learned to recite it from cover to cover. When he finished at *kuttab,* a boy went to Cairo to take an exam for admission to Al Azar University, which is a thousand years old, one of the oldest universities in the world.

My father was thirty-one when he finally got his degree in Arabic language, religion, literature, and poetry. Coming from a family of limited means, he was the first person in his village to be educated to this level. He became a sheik because of his knowledge of Islam and languages. It was education that made him somebody, and because of that he emphasized the value of education for his children. Before the revolution, when education was limited to people with money and social connections, he prayed to God for help to get *one* of his boys through high school—regular high school, not like the religious school he taught in, but a school where you could learn science and a great deal more.

Then came the revolution in 1952, which made education free, and all nine of us went on to get advanced degrees. Among us is a retired army general who also attended medical school; a Ph.D. from Harvard Law School who works as an advisor to President Mubarek; and another general who commands one of Egypt's major divisions. Then come me and my three sisters. One is a high school principal, one a chemist, and another an American-trained doctor working in Geneva. Finally there are the last two boys, a mechanical engineer in St. Louis and a banker in Cairo. So you see I'm not the only distinguished member of my family!

When did you first get interested in geology?

In the Boy Scouts during field trips. I also spent a lot of time walking out of Cairo into the mountains to the east and north of the city. One of them is called the Red Mountain, because it is made of beautiful red sandstone, and another is called the Yellow Mountain, because it contains pretty golden-colored shales and limestones. I would take my younger brothers out on weekend trips. We didn't have any money for public transportation, so we'd spend most of the day walking there and back, with a couple of hours remaining to make tea and explore the caves and mountains. I was fascinated by the variety of rocks, samples of which I'd bring back and label "red rock" or "yellow rock." But in high school I never heard the word "geology," and I was really more interested in medicine and anatomy. There was no question in my mind: I was going to be the best surgeon in the world.

When did geology enter the picture?

After the revolution the government decided who was going to get an advanced degree and in what. This depended solely on the needs of the country, whether at some particular moment it required dentists or engineers. My grades were sent to government officials who looked over all the kids graduating from high school. Back came an envelope giving me two choices: I could go to dental school or the school of science at Ain Shams University. I had no interest in becoming a dentist, so I took my papers to the registrar's office at Ain Shams and said, "What's cooking? What do you teach here?" That's when I heard the word "geology" for the first time. "What's geology?" I asked. "It has to do with people who go to mountains and collect rocks," said the registrar. "I want that one!" I said.

How did you get to the United States?

When I finished my master's degree in paleontology—I was studying the evolution of microfossils before a major die-off fifteen million years ago—my name appeared on a list of people being sent to the Soviet Union to get their Ph.Ds. My brother Assam, who had worked in the Ministry of Foreign Affairs and traveled many times to the Soviet Union, said, "You're not going to learn a damn thing from the Soviets. You shouldn't go." So I declined the offer, but a few months later there arrived a scholarship from the Ford Foundation to study in the United States.

The U.S. Bureau of Mines was supposed to assign me to a school. So I went to Washington and told them my brother Assam was going to Harvard and that, therefore, I should go to MIT. The man listened and then he said he was sending me to the Missouri School of Mines. "What's that," I asked. "It's the best school of mines in the country," he said. "It's where I went to school, and from what I know of your record and the needs of Egypt, it's where you should go. I'll make a deal with you. You go to Missouri for one semester, and if you don't like it, I'll move you to MIT."

So I traveled to Rolla, Missouri, in February 1960 to begin my first semester. The place was covered in snow, which I'd never seen before. I didn't know how to walk in the stuff. I didn't have any boots or warm clothing. I worked very hard so I could get out of there and go to MIT, but apparently I worked too hard. I called my advisor in Washington at the end of the semester. "Here are my grades," I said. "I'm leaving now."

"You can't leave," he said. "Your teachers tell me you're doing very well, and I'm not going to give you any money if you leave." It really turned out to be one of the best things that ever happened to me, because the education I got there was first class. When I later had the opportunity to study at MIT for a year, I have to confess I got straight A's for doing almost nothing.

After Missouri, your career takes a little twist when you go off to the Max Planck Institute in Germany to study meteorites.

My career is nothing but little twists! My first job offer after finishing my Ph.D. was to teach geology at the University of Heidelberg. My wife wanted to visit Europe. She spoke French, Italian, and Spanish, but neither of us knew a word of German. I learned it quickly enough, though, when I went on a six-week bus trip with thirty German students to visit all the mines in southern France and Spain.

My wife and I were going to spend a year-and-a-half in Germany and then move to Egypt. I started teaching her Arabic by writing words and phrases on little cards. I couldn't find any books on learning Arabic, so I wrote one, which was published by Dover Press. They paid me $350 for the rights forever, but in those days a dollar was worth four marks, so it was pretty good money. Anyway, it was enough to buy a refrigerator that we took to Egypt.

After rubbing shoulders with some of the world's great geologists and working on meteorite mineralogy with the top person in the field, I thought, "Now I'm ready to build a school of economic geology in Egypt the likes of which doesn't exist anywhere in the world." I shipped four tons of rocks from Germany to Egypt, samples I'd collected throughout the United States, Mexico, Canada, and Europe. My wife still reminds me of the days when she was pregnant with our second daughter and we'd go to the University at night to wrap rocks in newspaper. I took out lifetime subscriptions to all the important journals and lifetime memberships in the professional societies. I was set to spend the rest of my life in Egypt.

So what happened?

I was willing to work in any university, but the ministry wanted me to teach chemistry at a technical school in Suez. The school didn't even offer a degree in geology! I said, "I'm not going to take the job." They said, "It's that job or nothing." This was during Nasser's regime, when Egypt was a police state. No one spoke his mind, and you didn't say no to the government. But I did, and they didn't like it. So I left in the middle of the year for a job at Bell Telephone Laboratories in Washington, which got me involved in the whole fabulous experience of the Apollo space flights.

What have you learned from working on pyramids, tombs, and other archaeological sites in Egypt?

Working on the tomb of Nefertiti and discovering the solar boat under the Great Pyramid were both remarkable experiences. It was wonderful to take the things I'd learned from the space program—the most sophisticated technology that currently exists—and apply them to studying one of the world's oldest civilizations, which is also my cultural heritage.

Bert Sakmann

✦ ✦ ✦

ON a glorious spring day in Heidelberg, Bert Sakmann bicycles to work along the banks of the river Neckar. He spins up to a red brick laboratory in this university town of baroque buildings, parks his bike, and bounds upstairs carrying a paper bag holding an apple and one banana—his lunch. Except for a break at the neighboring cafe, Sakmann and I will spend the day in his laboratory talking about brain and muscle cells. How do they communicate with each other? What physically happens when a brain moves a finger, or thinks a thought?

A living body is a giant network of chemical and electrical signals coursing in milliseconds from the central nervous system out through neurons to muscle cells and back. Cells "talk" to each other by means of sodium or potassium ions—charged particles that travel through cellular gates and channels like airplanes stacked up on a busy day at O'Hare. Today's knowledge about ion channels, synaptic transmission, intercellular signalling, and the other mechanisms by which brains and muscles communicate is due in good part to the work of Sakmann, a fifty-one-year-old scientist at Heidelberg's Max Planck Institute for Medical Research. No wonder he and I have so much to talk about.

Sakmann and fellow Swabian Erwin Neher—"We are members of the same tribe," Sakmann jokes about his longtime colleague—shared the Nobel Prize in medicine and physiology in 1991. The award recognized a fabulous month of research a decade earlier when Sakmann and Neher invented the patch clamp technique for studying ion channels. Their ear for listening to the brain and body talking to each other was a micro-thin glass pipette twenty-five thousand times smaller than a human hair. The trick was to spear functioning cells in the process of emitting signals, without killing them. Once speared, these cells could be bathed in ions and manipulated in the laboratory just as they are in the body. These methods for tuning into cellular signals are now standard practice for cell physiologists and brain researchers around the world.

Always working at the forefront of the field, Sakmann was present at the discovery of today's accepted models for transmitters, receptors, ion channels, and other means of cellular signalling. He collaborated on the recombinant DNA experiments that isolated the proteins responsible for making different kinds of channels. He worked on therapeutic cures for epilepsy, diabetes, and other diseases caused by malfunctioning channels. And lately, he has been studying brain cells to see if he can discover in their structure evidence of the higher brain functions we call "learning" and "reason."

Among this raft of accomplishments, the courtly Sakmann remains unflappable and unassuming. He never uses the first-person pronoun in his published writing, and he ascribes much of his seminal work to chance. On getting the telephone call from Stockholm, his first thought was, "Oh, what a lot of luck!"

✦ ✦ ✦

Before we begin talking, do you need to instruct your troops?

No. We do small-scale science here. That's one of the beauties of membrane biophysics. We do our experiments by hand. There are no simple recipes, like the protocols in molecular biology. Experimental skill is still important in this science. We record electrical signals and interpret them. But there is no straightforward interpretation of an electrical signal, so like artists, we draw pictures of what's going on in the synapse.

What are you working on now?

Something in the brain called the nMDA receptor. nMDA is the amino acid aspartate. Because it's thought to be important for higher brain functions, such as developmental changes and maybe even learning, we're trying to figure out the structure of its receptor. We're competing with other groups, so things are hectic at the moment.

Are you winning?

It's a friendly competition. In physiology, particularly the study of brain function, two or three groups seldom perform exactly the same experiment, and everybody has different views of how synapses work. It's not like in molecular biology, where the group that discovers the sequence first gets all the credit.

What do you hope to learn about the nMDA receptor?

It's important for looking at how synapses, or connections, are formed in the central nervous system. This is a very exciting field that started to flourish four or five years ago, when we learned how to characterize the molecular transitions of this particular channel. In the last six months, the genes for this receptor have been cloned, and now we can manipulate the structure. We can experiment with what happens when you put different versions of the receptor in different parts of the brain.

When did you know you were going to be a scientist?

From an early age I was fascinated by biological cybernetics, which became popular in Germany after the war. This took principles developed by the American mathematician Norbert Wiener—principles for understanding how to hit a moving target with a gun, for example—and applied them to understanding the function of animals. By analyzing the flight pattern of a beetle, you could predict the beetle's movements and figure out the underlying principles of how its brain works, and you could do this without knowing any anatomy. The brain was conceived as a bundle of sensors feeding into integrators with differentiated output. Our ultimate hope was to build a machine that would explain how the body works.

So humans were equated with machines?

Making machines and explaining animal behavior seemed to be the same thing. Now we know it's a lot more complicated. But in Germany after the war this was the prevailing view. When I was a boy I loved to construct airplanes and ships operated by remote control. It was assumed I would become an engineer, but then I got interested in biology. So here was a field in which I could do both things at the same time, engineering and animals.

Can human nature be reduced to a cybernetic model?

Some aspects of it, yes. When you first start out in science, you think you can explain everything, even higher brain function. This enthusiasm is due to ignorance. As you gain more insight, your goals become more modest. Now I would be very happy to figure out something as simple as how a synapse in the central nervous system works.

What did you hope to learn through cybernetics?

Psychology had no interest for me, because it seemed too descriptive and unrealistic. It had no experimental tools. I decided to study higher brain function—what in German is called *vernunft,* which means "rationality" or "reason"—by focusing on the problem of pattern recognition. This is not the highest brain function, but, still, it requires functioning brains, and you can perform experiments on them. How does a cat recognize a mouse, even when it's upside down?

Perception would be explained by tracing neurological pathways?

Yes. I thought we could understand the principles of pattern recognition. But technology today is still struggling with what I was working on twenty-five years ago. We looked at brain function in engineering terms, and at the same time we wanted to construct a machine capable of recognizing patterns.

When did you get interested in neurobiology?

I realized the brain was a mystery I wasn't going to solve. So I moved on to something simpler and kept moving in this direction. Neurobiology didn't exist at the time—the term comes from America. We had neurophysiology, which dealt with the brain as a whole, and biophysics, which studied the functioning of membranes and channels. Only later did we put them together.

Is your family scientific?

My great grandfather directed a psychiatric hospital, and both my grandfathers were doctors, but my father was a theater director in Stuttgart. He was interested in science, but he lived in a completely different world. I like theater, particularly modern playwrights like Bertolt Brecht. I was intrigued by his theory about educating people through theater, but I was never tempted to follow in my father's footsteps.

Why did you go to medical school?

I thought if something went wrong with science, I could fall back on medicine, but my heart wasn't in it. Halfway through my medical studies, I thought, "It's time to get involved!" So I picked up a copy of the journal *Biological Cybernetics* and looked up the editorial board. The editor-in-chief worked on fruit flies, and I didn't want to work on flies. But I wrote to everyone else on the board. I got positive responses from several people, including Bernard Creutzfeldt in Munich, who said he needed someone to finish a doctoral dissertation in cybernetics. I probably shouldn't say this, because it's so superficial, but the truth is, I had met a girl who lived in Munich—she later became my wife—and I thought, "Maybe I can combine these two things." Creutzfeldt turned out to have a very nice lab, but my going to Munich was not strictly scientific.

What did you do in Munich?

Creutzfeldt was working on pattern recognition in the visual system, and he was also collaborating with a group of engineers from the technical high school. While we neurophysiologists studied the visual system of the cat, the engineers tried to build a machine that would recognize patterns. I was given the task of figuring out the synaptic organization of those special cells in the central nervous system that recognize contours.

How do you know when a cell recognizes a contour, or any other stimulus?

You drill a hole into the skull of an animal that has been anesthetized. While the animal looks at a screen, a micro-electrode monitors the electrical activity in its cells. This activity is usually transformed into an acoustic signal, a *bzzzz, bzzzz* that indicates you've found a receptive field.

These techniques seem crude compared to your later research.

Yes, but I'm trying to tell you how your ambition becomes less and less. It's guided by what's technically feasible. You want to have things under control, and every time you do an experiment you'd like to get the same answer. I decided psychology wasn't going to work. Then my experiments with the visual system didn't work either. It was only when I found a system with repeatable results that I was happy.

Before you gave it up, what did you learn about vision?

You have to manipulate your electrode *into* the cell to record what's called a membrane potential. Since these are living animals—they have a heartbeat, they move, they breathe—I got recordings on very few cells. I would go into a brain cell, record a membrane potential, and usually within seconds or minutes, the cell would become leaky and die.

What are membrane potentials?

Membrane potentials are the means by which cells communicate with each other. They result from the flow of ions, or charged particles, from inside to outside the cell, or vice versa. A flux of potassium across a cell membrane—potassium being positively charged—generates a membrane potential of about a tenth of a volt. All nerve cells, muscle cells, and probably all the other cells in your body generate these action potentials. This difference in voltage between the inside and outside of the cell is *the* requirement for signalling in the nervous system. It produces the electrical signal, or action potential, that travels along nerves, and it is the frequency of these action potentials that encodes the information cells transmit to each other.

Do cells have other ways of talking to each other?

Cells also communicate by means of synaptic potentials. These don't travel along the nerve. They are local. But when enough of them are added together, they reach a certain threshold, and generate an action potential. If you want to move your foot, your brain makes a decision, which is the consequence of many synaptic activations of cells called motor neurons. These motor neurons generate action potentials that travel along the nerve, until you move your foot. From the moment you decide until you wiggle your toe takes less than a second, and it's all done electrically.

Why do cells have different ways of communicating with each other?

Our bodies generate two kinds of electrical signals. Action potentials are used for signalling along nerves over long distances. Synaptic potentials are used to integrate information coming into cells from different parts of the brain.

A one-second response time seems rather slow.

Carl Lewis could do it quicker, in a tenth of a second or less, although even he has been slowing down lately. If you think about alternatives for transferring information between different parts of the body, you have the blood system, which uses hormonal signals, or the lymph system. But these take minutes instead of seconds. They are orders of magnitude slower than cellular signalling.

Why do nerve signals travel at different frequencies?

Information is not contained in one action potential, but in the different rates or frequencies at which they are transmitted. This is called frequency encoding, but how it actually works is a complete mystery. Researchers give animals visual clues, count up the number of action potentials fired by these clues, and call this a "response." But we really have no idea how this information is integrated in the central nervous system.

For many years you stopped doing research on the brain. Why?

I went to a summer course in Italy where Bernard Katz gave the introductory lecture on nerve muscles and synapses. This lecture made me decide to leave the visual system. I realized higher brain function was too difficult for me to understand. I wanted to work on the basic mechanism of how information is transmitted, so I thought I should look at the synapse we know the most about: the neuromuscular junction. No longer would I ask questions about how brains function. I would just try to understand synaptic transmission.

What is the neuromuscular junction?

The synapse between the nerve and the muscle. This is where most of our concepts about synaptic transmission have been developed. It is *the* model synapse. The only thing the neuromuscular junction doesn't do is learn, although our ideas about this might change. In yesterday's *Science* magazine somebody reported seeing long-term depression in the neuromuscular junction. This electrical alteration is associated with learning.

What is a synapse?

It's where the electrical signal is transmitted from one nerve cell to the next. Let's say you decide to move your thumb. An excitation generated in your motor cortex travels down your spinal cord to a motor neuron. From the motor neuron, it travels down your arm and is finally transmitted to the muscle. You have a chain of elements, neurons and muscles, that have to be hooked up. This is done by chemical substances that diffuse across a narrow cleft in the synapse. Electrical excitation can't jump across the cleft. Instead the signal is transmitted by a chemical substance. A small hole is created through which ions move. This move is registered as a change in synaptic potential, in other words, as a message to move your thumb.

Why are electrical signals converted into chemical signals?

I really don't know. In my personal view, I think it's an easy way to generate different configurations of these various elements. A nerve cell has many inputs, and one way to tune them may be to create more or fewer synapses. These synapses are not fixed. They can be lost. They can be regenerated. Electrical transmission is relatively rigid, but chemical transmission is flexible in its wiring and can change quite quickly.

Why do we need flexible wiring?

A lot of people are wondering, "What is the cellular mechanism underlying learning?" There is evidence that learning is due to changes in synaptic transmission. This is a central line of research in neurobiology. The process seems to involve the release of more or less transmitter, with resulting changes in synaptic transmission.

Can you give me an example?

Researchers have found what looks like a simple form of learning in *Aplysia,* a sea snail. Changes in behavior have been traced to altered synaptic transmission in the ganglion cells. Another example involves something called long-term potentiation. Behavioral changes resulting from shocks to the brain can be traced to increased synaptic response.

Can I learn something by building up neurons?

Every possible mechanism has been invoked: more synapses, more transmitters, more receptors, greater sensitivity, more channels or changes in their structure. We have examples supporting every theory. I don't want to be mean, but it's funny how the same groups claim "proof" for one mechanism and then another and then both. Personally, I don't think there is *a* learning mechanism. There are many, and that's the excitement.

Could they all work together?

It's too early to say. People nowadays tend to over-interpret their findings. Neurobiology is a fascinating field, but it has become a bit like Disneyland. There is a lot of self-advertising, which is no good for science. It is by no means clear, for example, that long-term potentiation—electrical changes in the cell lasting minutes or even hours—reflects learning and memory. This is the best paradigm we have, but it may be completely unrelated to learning and memory. Let's just say the field is a bit overheated right now.

When will we know how the brain works?

If you are content with understanding how a synapse works, or how synaptic potentials change with use, we might know this in three years. But a proper understanding of learning? It's not within reach. Higher brain function is too complicated. We don't even know what signals to look at. Is it enough to say a particular synapse has increased its strength? Is this is the basis of memory or learning? I would say, "No, it's not. This is not good enough." The process almost certainly involves an ensemble of cells and changes in the interconnections between them—changes that we are not about to understand.

You seem to have lost your early optimism.

I am coming back into this field after twenty years of asking simpler questions. Now that I think I understand the neuromuscular synapse, I am curious to know whether similar principles govern synapses in the central nervous system. When I began working again in this area, one of the most surprising things was how little had changed. People are asking the same questions I was confronting twenty years ago. I sometimes think we shouldn't even ask these questions. Instead, we should learn more about the structure of the channels. But we do have new tools that may be bringing us closer to getting answers to our questions. Soon we will be able to understand changes in synaptic strength down at the molecular level, although again I have my doubts about whether this has anything to do with learning and memory. Still, it's a lot of fun to find out!

What questions are you asking about the central nervous system?

We are trying to learn the rules that govern synaptic transmission in the visual cortex. How much transmitter is released? How many receptors are activated? What determines the shape of the synaptic current? A muscle is a simple structure. It's just a tube with one contact. But a nerve cell in the central nervous system is a lot more complicated. Along with a cell body, it also has a very extensive dendritic tree, which consists of big receivers covered with synapses. These synapses integrate information from various inputs, but their location on the dendritic tree also plays a role in learning.

Are you experimenting on rats and cats?

Rats. This reminds me of my very beginnings in science. These questions were always in the back of my mind, but to answer them I needed new methods. Three or four years ago, when I decided to work again on neurons in the central nervous system, I developed a brain slice technique that's quite helpful for looking at synaptic channels. It's too much to ask what the channels are good for, but at least we can find out how often they open and close and which ions are passing through them.

When did you develop your brain slice technique?

In 1988. It was dictated by my renewed interest in the central nervous system. I had worked previously with cultured neurons. But cells in tissue culture lose their specific properties. The only alternative is to work with living brains.

How does the technique work?

We cut a slice of brain three hundred microns thick. We remove some of the overlying tissue, called neuropile, with something that looks like a little vacuum cleaner. Using certain optical tricks we can look through the slice and see different cell types and their dendrites, which we then explore. Once the cell body is exposed, we record its signals with the patch clamp technique.

What does the patch clamp technique actually record?

The flow of ions, or charged particles. The nervous system runs on sodium ions, potassium ions, and chloride ions. If sodium and potassium move in, they excite the synapse. If chloride ions flow in, they inhibit it. The axon has a threshold of about –55 millivolts. So anything that drives the membrane potential close to –55 is excitatory, whereas anything that increases the inflow of chloride is inhibitory.

Why would I want to inhibit synaptic transmission?

You have to control excitability in your brain, or it becomes epileptic. One way to treat epilepsy is to increase inhibition. Inhibition is everywhere, in every cell. You have so much excitation coming in from your environment that you have to reduce it and concentrate only on the information that's important.

Without cellular inhibition I would become insane?

You would have seizures. Inhibition is as important as excitation. Valium, for example, makes you less excitable by increasing inhibition in the central nervous system.

How fast do synapses fire?

About a thousand times per second, so every millisecond you can have an action potential, or a signal.

What do ion channels look like?

A funnel and a gate large enough to pass one ion at a time. The inner wall of the channel is made up of what look like five barrel staves, which tilt in to close the channel. When the transmitter binds to the receptor, the staves untilt and the channel opens. At least this is one theory. The actual mechanism remains a mystery. Their is no X-ray structure available, so we really have no idea what a gate looks like. We have a better idea about the protein that forms the inner walls of the channel. Its amino acids decide which ions are allowed to pass. This is the other type of research I'm doing, which is just as intensive as looking at synapses in the central nervous system. Using recombinant DNA techniques, I'm trying to discover how these amino acids function.

Can you see ion channels opening and closing?

No, we can't see the channel. We can only measure the current flowing through it.

How do you know you are looking at a single channel?

In a single patch of membrane you'll have several different kinds of channels. Some are big, some small. Some gate fast, others more slowly. You have to use various tricks, like recombinant DNA, if you want to distinguish between these channels. Let my try to explain why this research is so exciting right now. In native tissue, you have voltage activity channels and all sorts of other channels. The genes for almost all these channels are now known. You can put these genes into cells that will make channels of a particular kind. You can select cells that have very few channels of their own, like the egg cells of the *Xenopus* frog, and get them to express any channel you want. Channels are made from subunits. These subunits are coded for by genes. These genes have been isolated. Now we put these genes into other cells that have few endogenous channels of their own. What results is a very controlled, very neat way of studying single ion channels.

Can they be transferred from one species to another?

Yes, and this transfer is permanent. We can put the nMDA receptor from a rat brain into the eggs of a frog. You take the genes, make RNA, and inject it into a frog's egg to get an oocyte stuffed full of nMDA receptors from a rat. This allows you to manipulate the channels, which is how we discovered that ion channels are made from different subunit proteins. They are put together from building blocks, like a game of Lego.

Why are ion channels constructed like Legos?

I think it's a convenient way to adapt channels to particular needs. In the next few years, by replacing channel subunits in transgenic animals, we will be able to find out why there is such diversity.

How often do we replace Lego units?

In the early stages of mammalian development, one of the subunits in the acetylcholine receptor disappears. It's replaced by a channel with another subunit and different properties. Now I want to see if this switching of channel subunits takes place in the central nervous system. This is the research I'm doing right now, looking at brain cells to determine their functional properties. We've found that they *do* vary in function, so now we're trying to figure out the molecular basis for this variance.

How do we alter our brain cells?

Brain cells are not fixed entities. They are continuously changing their repertoire of channels. During development, for example, you see drastic changes in the kinds of receptors in channels. How these changes alter the function of the synapse is one of the things we are trying to determine. I think the real function will be visible only when we are able to knock out particular subunits, and that's one line of research we are now following. But, again, what I am describing is highly speculative. The detailed physiology of the central nervous system only began two or three years ago.

How do brain receptors change?

All the receptors in our brain have a metabolic lifetime. They are built, put in the membrane, and are then taken away again to be metabolized.

We get new ones. It seems these receptor channels are not assembled in the membrane where they finally end up. They are put together in an intracellular compartment, a vesicle, and this vesicle is then transported to the membrane, where the receptors are incorporated. Most of our ideas about how this process works derive from earlier research on the neuromuscular junction. This is a marvelous model that we will build on in trying to understand the central nervous system.

What are knock-out experiments?

Let's say you suspect the nMDA receptor is important for learning. You would expect that knocking out the nMDA receptor would strongly affect an animal's ability to learn. This is a straightforward prediction, and it's what we and a lot of other people are trying to do. I suspect redundancy, though. There are subunits that can replace each other without any obvious change in function. Knock-out experiments are very expensive and very time consuming; so every discovery is a struggle.

When you're studying brain slices, are they still "thinking"?

The brain slice rests in an oxygenated solution, and the synapses are at a low enough temperature that things don't fall apart very quickly. You keep it at –25 degrees Celsius, and under these conditions the functions we want to look at are well preserved, for example ion transport and communication between different parts of the brain. This lasts for six or seven hours.

Early in your career you moved to England. Why?

Before doing fancy experiments with higher brain function, I realized I should learn about simpler things, like the basic mechanism of synaptic transmission. So I joined Bernard Katz's lab in London. Katz and Ricardo Miledi had just discovered what they called "membrane noise," from which they derived an estimate of the elementary event. This is the current that flows through a single channel when it is opening or closing. At the same time, a Taiwanese scientist gave Miledi a snake toxin called *bungra,* which specifically affects the acetylcholine receptor at the neuromuscular junction. It was then possible for the first time to count the number of receptors and channels in a synapse.

It was in Katz's laboratory, in the short space of two years, that the biochemistry and physiology of the synapse became molecular. We took part in all this, and it was terribly exciting to sit in on the everyday discussions of what was happening. It now became crystal clear what I wanted to do: look at the molecular properties of synaptic transmission and ion channels in particular. I had the electrical signals, and now I also had a biochemical handle that allowed me to label the acetylcholine receptor. The next step was to extract and purify and finally crystallize these receptors, which would allow us to study both their structure and function. I was very fortunate to be working with Katz when all this happened.

What year was this?

Katz and Miledi discovered membrane noise in 1970. At the same time Miledi began his experiments with the *bungra* snake toxin. He labeled the toxin radioactively and then counted the number of receptors with which it interacted. Suddenly, the postsynaptic membrane became translucent. We could see the elementary event and calculate how many receptors were involved. This inaugurated the molecular analysis of synaptic transmission.

A synapse has how many receptors?

About ten million.

Why so many?

You need a lot of current, at least at the neuromuscular junction, and you have spare synapses that aren't used.

How often are they replaced?

It depends on the animal and its age. In young rats, they are replaced every other day. In adults it's every fortnight. If you cut the nerve they turn over faster. So it's a nerve signal that regulates this development.

Did Katz get his Nobel Prize for this work?

No, he received it for his work on vesicles. He showed that transmitter is released in packets, or quanta. He did this research in the 1950s, and he was already a famous man when I went to work in his lab. I was lucky he took me on.

How much transmitter is required to move a muscle, or think a thought?

Signals at the neuromuscular junction release about three hundred packets, and each packet of transmitter activates about a thousand channels. Many fewer are used in the brain, and each packet activates from five to twenty channels. Why is this? It's one of the questions we're working on. I think it's because the brain is not simply relaying information. It is also integrating information. Each brain cell has thousands of inputs, and the smaller the effect, the more finely tuned your response.

What do you think of today's research on "neural nets"?

It's like my early work on cybernetics. It's trying to build a hard-drive model, made up of elements that behave like neurons, that will perform tasks similar to those undertaken by the nervous system, such as pattern recognition. Now it's called "neural nets." Twenty years ago it was called "homogeneous layers." It's big business, and the guys doing it are so clever they don't have to wait until physiologists have figured out how the visual system actually works. They just do it the way they think it *should* work, and they're very successful. But whether this reflects what's going on in the central nervous system, I doubt.

I notice you don't wear a watch. Do you have an innate sense of time, or don't you care what time it is?

Both. A watch makes me nervous, and in the lab I don't want to be nervous. Fortunately for most of my life I didn't have to take care of time. My family was always very generous. I could leave the lab whenever I wanted without people being offended.

How has winning the Nobel Prize changed your life?

What I found most difficult was giving popular lectures. The first few were a real disaster. It took a long time to figure out how simple I had to be. The toughest thing was this notion of currents and membranes and voltage. It's very difficult to explain how the current is carried by ions and not by electrons. The way the cell signals and how messages are sent by action potentials or impulses that travel down the axon is hard for people to grasp.

How does your work differ from Erwin Neher's?

His big thing is secretion—the release of vesicles—while I focus more on channels and tools for studying them with molecular biology, which I spent of lot of time doing in the mid-1980s.

You don't use the word "I" in your articles.

No, I hate it. Fifty years ago it was appropriate to say "I." But nowadays you're building on so many experiments done before you, and you're using so many techniques and concepts about whose origins you are ignorant—because they are textbook knowledge—that it's very easy to put the "I" in the wrong place and make it sound as if you invented everything. Katz, for example, never used "I." He would say "several lines of evidence suggest" or "the experiments show clearly," but never, "I discovered. . ."

How did you meet Erwin Neher?

Before going to work in Katz's lab, I thought I should know something about voltage clamping, so I spent part of a year learning the technique from Erwin. Voltage clamping is the way to record channel currents. Erwin had just started at the Max Planck Institute, where he was using pipettes to record from different parts of the neuron. He was trying to find out if these different parts had different types of channels. Anyway, he and I got along quite well.

What was his background?

He came from a technical high school. He was recording currents from snails, which are very convenient to study because they have big neurons, up to a millimeter in diameter. These huge cells are easy to penetrate, and after my frustrating experience with the central nervous system, it was nice to do some easy experiments for a change. I had finished my thesis and married and was experimenting with Erwin's snails for a few months before going to London.

This was the start of a long collaboration?

Erwin and I are very sympathetic, because he's also a Swabian, from southern Germany. We speak a similar dialect. We are members of the same tribe. I was very lucky to meet him, because he showed me how to prepare the animal, and he taught me electronics. We had a very good time going down to the electronics workshop and building an amplifier together. He had just written a little book for medical students about using electronics for electrophysiology, which was very helpful.

Rather than spending hours preparing live animals and trying endless times to penetrate their brain cells and stay intracellular—which lasts only two or three minutes until everything disappears—with voltage clamping you could really concentrate on the experiment. Putting electrodes into Erwin's snail cells took a few minutes, and then you could play with the membrane, and that's what I really wanted, to have things under control. That confirmed my decision to leave the central nervous system and work on what might be less exciting questions, but with more precision.

Where did you go after London?

Creutzfeldt invited me to accompany him when he moved from Munich to Göttingen, which has another Max Planck institute that does a lot of physical and chemical research. This was exactly the kind of environment I wanted. It had everything—lasers, physics, chemistry, cell biology, molecular biology—and they were also very strong in physical instrumentation. I said, "I'm not going to work on the central nervous system. I want to work on synapses and channels. Maybe after a few years, I'll move on to the retina. But for the moment I want to concentrate on really simple things."

How did you and Neher get back together?

When I visited Göttingen, I found Erwin there. He was part of an entire department devoted to studying artificial channels. He had switched from biological to artificial membranes at the same I began researching something more handleable than the brain. I said it would be nice if we could work together, and suddenly it occurred to us that we might be able to adapt artificial membrane techniques to studying biological membranes. Erwin wanted to get back to physiology, and I wanted to use the modern techniques I had learned in London, so we agreed to work together. We would use Katz's noise technique to estimate the size of muscle fiber receptor currents and then study them more closely with pipettes. But we couldn't get Erwin's pipettes to work on biological membranes. The seal wasn't tight enough. Fortunately, I had developed a new technique for cleaning membranes, and after we found it worked, all the subsequent steps were quite obvious.

Who first used pipettes to study cellular activity?

They have been around since the 1940s, but no one back then ever dreamed the technique would be used for looking at single channels. To reduce the background noise, we calculated we needed a pipette with a diameter of two or three microns. Today we have pipettes down to 1/25,000th the size of a human hair. If our estimate of the size of the elementary current was correct—which we didn't know, since no one had ever actually seen one before—then our pipettes should have been able to record currents in the shape of little blips. It was terribly exciting when we finally saw them.

How did the experiment work?

You press the pipette onto a nerve cell, and the more you press—without puncturing the cell—the better your seal. Just by pressing and using clean surfaces we were able to record these blips. This was very nice, but Erwin had arranged to go the States for a year, which happened just when we were trying to set up a few controlled experiments. I was struggling along on my own, and he was struggling in America, when one day he called and said he had found much nicer channels in *Rana pipiens,* an American frog. *Pipiens* has longer channels than the European frog and larger currents. This helped us do a few experiments. But we still weren't quite sure what we were measuring. Then I got hold of some *pipiens* in Germany, and this made a big difference.

Erwin came back for Christmas and we wrote one of our early papers. In the meantime, I had been working on the rat, trying to repeat our experiments on a different preparation. The noise analysis indicated rats also have large channels. Basically, we spent the next five years improving our techniques. We made programs for analyzing the recordings. We perfected the electronics. Erwin built different amplifiers, while I worked on new preparations. One thing that drove us crazy was the preparation of these muscle fibers. It was unspeakably boring, but we had to do it, because we were getting a lot of results and publications.

How did you prepare your samples?

It took two or three hours. We had to dissect the frogs and clean them with enzymes, at which point they became rather fragile. It wasn't at all fun. I was really getting fed up, when one day a nice American guest came to visit, Fred Sachs. He's a physicist, but he had been working on membrane preparations, and he showed me how to prepare muscle fibers from embryonic rats or chicks. You prepare them once, and then you let them grow, which gives you a preparation for the whole week. This opened the way to trying out many different experiments.

But we still had a major drawback in not getting tight seals. It helped to suck a piece of the membrane into the pipette, and we had been cleaning them by dipping their tips in resin. But one day Erwin was feeling lazy, or maybe he was in a hurry, and he forgot to clean the pipette. He just used a freshly pulled pipette and sucked on it, and suddenly he found a huge increase in seal resistance. Just as suddenly it didn't work anymore. It was a mystery. Sometimes it worked, and sometimes it didn't. Then we discovered that the pipette catches dirt when it enters the cell culture, and if this dirt attaches to the tip, it doesn't work. For a month or two we were desperate. Then one of the postdocs discovered a simple trick. A little puff of air through the pipette solved this problem. Everything was working again!

What happened next?

I discovered, again by chance, that if you have a tight seal, you can remove a patch of cell and still get recordings. I accidentally knocked the table. The pipette leaped out of the dish. Here was my preparation. Here was my pipette up in the air. And I was still recording channels! "What's going on?" I thought. Then I realized I had removed a patch of membrane from the cell with its seal intact, and again this was a remarkably lucky observation. Now I could manipulate the ion bath on *both* sides of the membrane. This was the perfect experiment for researching ion flow through a channel.

The word "luck" appears frequently in your papers.

Yes, it does. When I got that telephone call from Stockholm, my first thought was, "Such a lot of luck!"

What did you do with these excised cellular patches?

We discovered another new technique that came just from increasing suction on the pipette. If you keep sucking while still maintaining a high resistance seal, you can gain access to the *intra*cellular side of the membrane. With conventional electrodes, you basically destroy the patch and the membrane as the pipette moves back and forth over the cell. But in this new technique, the glass becomes a continuation of the membrane, which means it doesn't leak. Then we discovered another way to remove the pipette and end up with a sealed-off vesicle. This gave us what we called an outside-out patch, with the membrane oriented in such a way that we could bathe the *outside* of it.

When did this happen?

In 1980, in two busy weeks after Christmas. We were very happy! Erwin and I and two very dedicated postdoctoral students from Australia and France worked like hell. Fred Sigworth, an electrical engineer from Yale, did all the electronics. He built one apparatus after another, making it easier and easier to record signals. We worked day and night and had a lot of fun. Erwin's and my labs were adjacent, and we had no doors. We were just yelling, "I have a new configuration!" Each of us still had other problems to work on. But then we realized the methods we had just discovered were too exciting to be left alone.

With two Germans, an American, a Frenchman, and an Australian in your lab, what language did you speak?

English.

But with occasional exclamations in German and French?

Yes, "Help!" As the news spread around, many visitors came from all over the world bringing their own preparations, until we became a bit wary and decided to write our methods paper. It was sort of a cookbook. Then we had the idea to collect all the tricks of the trade and offer a course to our friends and collaborators. Finally, after writing a book about it, we got back to our individual problems.

How widely used is the patch clamp technique?

Practically every cell physiology lab in the world uses it, and now these techniques are spreading into research on the central nervous system. They revolutionized the study of cellular communication. First of all, they enabled us to look at the elementary event. We could study current flowing through single ion channels as they open and close. This brought new insights into the gating process. The same thing happened for the conduction process. This was particularly helpful for studying channels created by recombinant DNA, which was another field that was revolutionized by the new technology. Because the whole intracellular gating machinery has become available, this has opened up the study of mammalian cells, which previously could be investigated only with sharp electrodes and great difficulty. Now there is not a single cell you can't characterize with respect to its currents.

What have you learned about cells?

They have many types of channels in their surface membrane, which we never expected. This goes for plant cells as well as muscle cells, nerve cells, secretory cells. Plant cells are packed with these channels, swelling, opening, serving some sort of function. This is another aspect of the revolution—the realization that all cells have many types of channels in the membrane.

What diseases can be cured by this research?

The standard example is diabetes, where secretion depends on the function of a particular channel. This, too, is part of the revolution. Many new channels have been discovered, a lot of them with immediate functional significance for particular diseases. This includes heart and muscle diseases, diabetes, cystic fibrosis, and epilepsy. The patch clamp technique won't cure these diseases. It just allows us to figure out what's wrong. In cystic fibrosis, it's a malfunction of the chloride channel. In diabetes, it's a malfunction of the potassium channel.

How does a channel malfunction?

In cystic fibrosis, the chloride channel doesn't open the way it should. A protein malfunctions, but the molecular details are unclear. Five years ago we would have been content with this much knowledge, but now we want to know more. Is there a problem with ion conductance or gating or docking of the metabolite? In the case of diabetes, our research is centered on one step in the cascade from cellular signal to the secretion of insulin. We have found the site of action, but the cause of the malfunction still eludes us. Finding the site of action for sedatives like Valium or substances that may prevent a stroke or epilepsy is another type of research I'm involved in. Epilepsy is a hyperexcitability. It can be cured by increasing inhibition or decreasing excitation. To do this we are developing drugs that act on the glutamate receptor. You plug the hole, basically, to inhibit glutamate from binding to the receptor.

Are drug companies using these ideas?

Bayer is spending a lot of money developing calcium channel blockers and researching other drugs that increase calcium flow. If the heart is not contracting properly, this can be corrected by opening calcium channels. How these drugs work is illuminated by ion channel techniques. They clarify the signalling pathways and allow you to think about how to interfere. I'm collaborating with another drug company on a cure for epilepsy. After I discover the mode of action and site of action for these drugs, the company will develop them.

Do you hold any patents?

No. It's too exciting, and too much trouble. We have so many things to learn, this is really a distraction. What's the use of filing for patents? I'm not going to make money out of science. It's far too interesting for that.

Why did you collaborate for several years with a molecular biologist in Japan?

I was trying to figure out how channels work using recombinant DNA. This research was done with the late Professor Shosaka Numa from Kyoto. He did the molecular biology. We did the functional analysis and interpretation. We hoped to derive rules for the broad class of channels important for synaptic transmission. We haven't got that far, but this is a line of research I still continue.

Why did you work with a laboratory in Japan?

In the 1980s there was another revolution in neurobiology that was more or less initiated by Shosaka Numa. He applied recombinant DNA to the study of ion channels. He used techniques developed for other proteins to isolate the genes for channels. He found that transmitter-gated channels serving synaptic potentials, voltage-gated channels signalling along the nerve, and second messenger channels—these are the three classes of channels, which are differentiated by the way they operate—all had similar amino acid sequences.

What are second messenger channels?

They are called second messenger channels because they involve a receptor, an intermediary substance called a G protein, and the channel. It's a much more complicated signalling event, linked somehow to the intercellular metabolism. Second messenger channels are responsible for slow changes in excitability—slow meaning hundreds of milliseconds or seconds. Like hormones, they change the excitability of neurons over longer periods of time.

Will you find more kinds of ion channels?

I expect that many different kinds of channels have yet to be discovered. For example, there are channels that are operated by stretch. You stretch the membrane and the channel opens. But I think we have found the three major families of channels responsible for quick signalling. This was the work of Numa. He characterized the amino acid sequences in channels and realized they form families. It was then possible to manipulate these amino acid sequences and change the properties of channels. Now we want to crystallize a channel and see its 3-D atomic structure, although so far this has not been possible.

Are you looking for a "unified field theory" of ion channels?

That would be too much to ask. But all the families we have looked at so far have similar energy profiles, and they share a common structure in their subunits. So it's not completely hopeless. A coherent picture is emerging, but it's a bit too early.

Are receptors in muscle tissue the same as those in the brain?

Yes, although brain receptors have additional properties. Let me give you two examples. Inhibitory channels in the central nervous system can be modulated by sedatives like Valium. Does this mean there is endogenous Valium in the nervous system? Why else would it already have channels for Valium? You don't have these channels at the neuromuscular junction. The same thing applies to nMDA receptors. In the presence of glycine the current through these receptors is twice as high. These features are particular to the brain.

These exceptions aside, do brain receptors function like those in muscle fiber?

I think so. A lot of people ask me this question and then make fun of the answer, accusing me of saying that the brain is nothing more than a big muscle. I would never say that. But I think there are common principles for all transmitter-gated channels that subserve fast synaptic transmission.

If the brain works like a muscle, then I should be able to exercise it.

[Laughing]. It would be nice to improve synaptic functioning, but I have no idea what it would take. That's what we're trying to find out.

Jonathan Mann

✦ ✦ ✦

As word spread around headquarters of the World Health Organization that Jonathan Mann was quitting as director of the Global Program on AIDS, a hundred staff members gathered in the hallway to give him a cheering ovation. He acknowledged their support with a clenched-fist salute.

"He's on the warpath. The bullets are flying up and down the corridors," said a GPA staff member after Mann went public with his reasons for quitting. Although he had announced he would step down in June 1990, Mann was forced to vacate his office two months early.

Mann and a single staff member had founded the AIDS program in 1986. Four years later, it had grown into a $109 million effort with close to three hundred full- and part-time employees, making it WHO's largest program. As Mann traveled the world, implementing a global strategy for fighting the epidemic and helping to establish one hundred fifty-five national AIDS programs, he became an eloquent advocate for human rights.

No discrimination, no stigmatization, no coercive policies will stop the spread of the disease, he says. Voluntary testing, confidentiality, and respect for the rights of people with AIDS are the only public health measures that will work. Mann, who is equally proficient in French and English, can be credited with stopping the imposition of mandatory AIDS testing and border controls in numerous countries. By sheer force of personality he became chief medical officer to the globe on AIDS-related issues, and as Mann sees it, there is very little in the world—from the fall of the Berlin Wall to the democracy movement in China—that is not related to AIDS.

The forty-six-year-old Mann is way ahead of most people in his thinking about the human rights and political dimensions of AIDS. One of the people whom he long ago outdistanced is his former boss, the Japanese neuropsychiatrist Hiroshi Nakajima, who in 1988 was the first non-Westerner to be appointed director-general of WHO. Nakajima fits all too nicely into the villain's role in this story. He embodies what AIDS activists call the conspiracy of silence, although in this case, the conspiracy is quite active. Believing that AIDS gets too much attention compared to other diseases, Nakajima cut the human rights and public relations people out of Mann's staff, and he blocked new appointments to the program. "He put up an administrative wall against us that completely paralyzed our efforts," says Mann.

Mann felt strongly that WHO should play a leading role in guaranteeing equal access, for rich and poor alike, to expensive AIDS medications and future vaccines. "A hundred times I begged the director-general to act on these issues. There was never, ever any reaction." But for Mann the most crippling silence lay in Nakajima's refusal to speak out against AIDS-based discrimination. "Human rights is not on Nakajima's agenda," says a former GPA staff member.

After resigning, Mann went on record calling Nakajima unworthy of his job and a disgrace to the medical profession, which is grappling with a virus that has already infected ten million people worldwide. To date, over two million HIV-infected adults have developed AIDS, and most of them have died. By the year 2,000, the total number of infected people will have risen to forty million.

The World Health Organization's headquarters in Geneva occupy a massive glass cube that sits on a hillside overlooking Lac Léman. Filled with doctors advising doctors, this United Nations agency functions as the world's consulting room. Its corridors rival the Pentagon's in unrelieved length and anonymity. At the end of one of these corridors is Mann's office, so crammed full of researchers and computers that the only spare chairs for visitors are kept in the hall.

Several weeks before his resignation, I spent four evenings with Mann watching the light fade on Mont Blanc and talking to him about AIDS, primarily the cultural and political dimensions of the epidemic. Mann, who wears bow ties and works with his shirt sleeves rolled up, combines the serious mien of an intellectual with the tousled look of someone in a hurry. Sitting under a reproduction of Botticelli's Venus, we tried to comprehend an epidemic that Mann considers one of the fundamental revolutionary events of the twentieth century, and perhaps the twenty-first century, as well.

Brought up in a Conservative Jewish family, the eldest son of a Boston psychoanalyst father and social worker mother, Mann studied art history at Harvard before going to Washington University for medical school and then back to Harvard for a master's degree in public health. He worked for two years as an epidemic intelligence officer at the Centers for Disease Control. This was followed by a seven-year stint as chief medical officer for the state of New Mexico, before he returned to the CDC in 1984 to become assistant to the director of the AIDS program.

Mann was dispatched to Kinshasa, Zaire, to establish *Projet SIDA* (SIDA is the French acronym for AIDS). This was the first comprehensive study of AIDS in Africa, and it provided the original data on the incidence and spread of the disease in Central Africa. With only three expatriates on the staff, Mann's project was also a model for how to train Third World scientists into the cadre of international researchers. Mann left Kinshasa for Geneva in 1986. Currently director of the Harvard AIDS Institute and a professor of epidemiology and international health in the Harvard School of Public Health, he remains a committed activist in fighting this epidemic, call it AIDS or ignorance.

✦ ✦ ✦

Tell me about your medical training and early work on AIDS.

The one book that most deeply influenced my choice of career was Albert Camus's *The Plague*. The hero of the book, Dr. Rieux, exemplifies for me a special kind of humility and unwillingness to yield in front of the inescapable fact of death. Reading Camus, as much as anything else, is why I decided to go to medical school.

Have you ever dealt with plague?

I encountered many cases of plague, which is caused by the bacillus *Yersinia pestis,* during the ten years I worked as chief medical officer in New Mexico. I was particularly interested in the social and cultural features of the disease. Plague is treatable if it's caught in time. There's no reason in the modern world to fear it, but it's one of those diseases that carry visceral connotations. Anxiety about the disease could cause as much damage as the bacteria itself. When you tell somebody they have plague, you see a ghost take possession of them. They think immediately of the Black Death.

I may be inventing connections in time, but it seems to me that I got into medicine because of my feelings about Camus's book on the plague, and that led me to working on plague in New Mexico. Then I began dealing with AIDS, which has been called the plague of the late twentieth century.

Is AIDS the plague of the twentieth century?

Diseases aren't just diseases. They aren't simply microbes and immune systems. Diseases have *meaning* for people, especially infectious diseases. This is why it's odd that epidemics, even those that have caused tremendous destruction, have left behind curiously little artistic or literary production. You'd think the Black Death, which killed more than a third of the European population in the mid 1300s, would have made a mark on every aspect of human life. But frankly, I think it's more remarkable for the *absence* of that mark.

What sort of cultural mark—or absence—is AIDS going to make?

AIDS is the first truly global disease because it's *felt* to be a global disease. You might say smallpox was a global disease, but it wasn't. When the smallpox eradication program began, smallpox was already gone from many parts of the world. AIDS has been around for awhile—we don't know how long—but the current epidemic is new, and it's taking place in a world so tightly linked together that you can go to the most remote village anywhere and find evidence of the global consumer society.

If you draw the curve of the number of international travelers, it's an epidemic. Tourism is becoming one of the world's largest industries. This level of movement and exchange is far beyond anything that has ever existed before, and this makes us more vulnerable. People realize that our world today is different from the world in the past, and it's not just because it's the world we're living in. It's different because this issue of international exchange and movement and globalism is actually part of an emerging reality and consciousness.

What is this emerging consciousness?

AIDS doesn't *have* to be more of a global issue than nuclear war or concern about the environment, but it is, and that's a remarkable thing. There are no other global problems perceived as such. Hunger, poverty—these are certainly global problems, but too often they feel like local problems or somebody else's problem. On the first World AIDS Day in

1988 every single country in the world, without exception, carried out some activity against AIDS.

Some countries initially felt they were isolated from AIDS, but now they realize there is no such thing. There is no border, no boundary. We've learned that walls endanger because they encourage a false sense of security. Even if you could impose the perfect program for screening international travellers, infected people will get in, and some of your own citizens will come back infected. In the meantime, people won't follow safer sex practices, because they figure they're protected inside their walls.

When I was in China a few years ago, my hosts took me to visit the Great Wall. Standing in its shadow, I asked them, "Did it work?" "No," they said. "The invasion came from the other direction." This is typical. The idea that national boundaries are transcended by disease is now part of our understanding of how artificial these boundaries are. AIDS may be the harbinger of a new way of thinking about the world. This isn't going to be the last global epidemic, but it may be the first one to give us some real insight into how to think globally about disease.

It seems tragic that Marshall McLuhan's "global village" is being engendered through a sexually transmitted disease.

I don't see sexually transmitted diseases as particularly tragic. What they may help us understand better is our sexuality. One of the major barriers we face in trying to get countries to deal effectively with AIDS is the tremendous gap between social myth and social reality. I think closing this gap is a useful step. It's important to deal with things as they are, and not as somebody would like them to be. In that sense there's something noble about the way this disease is engendering knowledge of the self.

When I speak to national ministers of health about prostitution, drug use, homosexuality, child abuse, and related subjects—all of which have direct relevance to people's vulnerability to HIV infection—I often get the official line. "There are no homosexuals in our society. Oh, maybe a few, but they're all foreigners. We have very few prostitutes, and the only ones we have come from the neighboring country. Child abuse? We're not like the United States or the United Kingdom!"—two countries that have published data on child abuse. "As for promiscuity? No, there's none of that. Just some young people tainted by foreign ideas." The paradigm of the foreigner as a source of impurity, often sexual impurity, is a drama you see played out all over the world. The reality, of course, is that sexual practices are often radically different from what the myth describes.

I have asked a lot of government officials and experts, "At what age do young men and women in your society begin to have sexual intercourse?" This is not prurient curiosity on my part. I'm trying to figure out when you might start certain kinds of educational programs. The expert thinks a minute. He may take on a reflective look as he considers his own adolescence, and then he makes a decision. Is he going to tell me when he had sexual intercourse, or when he thinks he *should* have had sexual intercourse? The answers are not at all scientific, and frankly, people often don't know what's going on in their own society in terms of sexual practices and drug use.

What is AIDS teaching us about our sexuality?

The AIDS epidemic is composed of a virus and people and behavior. We know more about the AIDS virus than any other virus, but we know relatively little about the behaviors involved in spreading it. We have learned a couple of important things, though. The need to protect the rights of infected people and our understanding of the medically harmful effects of discrimination are some of the major public health discoveries of the 1980s. Discrimination and marginalization are themselves dangers to public health. They have to be fought to protect human rights, but also for the pragmatic, public health-based reason that a disease like AIDS can be controlled more successfully through respecting people's rights and dignity than through punitive or coercive approaches.

This critical lesson has relevance to our understanding of globalism and the best way to deal with threats to our common security. We're seeing plenty of evidence in the world right now that coercion may produce a temporary period without conflict, but the conflict is seething below, and coercion in the end never represses it. The symbolism of the fall of the Berlin Wall is really quite important. Which side of the wall crumbled? It fell from the inside. It was another wall creating vulnerability and danger for the people within it.

The AIDS epidemic seems particularly tragic for young people. An exploratory phase in life now has to deal with the added fear of death.

This is the first generation in the history of the world to grow up with AIDS, although it will not be the last. Young people have to figure out how to develop a capacity for intimacy, including sexual intimacy, without getting AIDS and without fear dominating that intimacy. Learning how to love in a world with AIDS is a lot tougher than anything we had to face when we were growing up; so I'm not sure we have a lot of useful guidance to give young people. That's why it's important that youth educate youth. In my experience, they don't disappoint you if you give them the opportunity.

One of the critical principles in health education is that the message be developed by and for the people who are going to receive it. I keep on my desk this brochure, which is the ugliest piece of literature I've ever seen on AIDS. It was designed by a group of French intravenous drug users. It reminds me that it doesn't make a bit of difference what *I* think about the way the message is presented. The only thing that counts is what *they* think about it. We often fall into the trap of thinking we know what other people need. But who are we kidding?

The sexual revolution of the Sixties existed for only a brief moment in history. Was AIDS the major cause for its demise?

Even during the sexual revolution, people were predicting it would be followed by a sexual counterrevolution. You might remember the chilling effect from herpes, which came before AIDS. You could have sexual intercourse with somebody who looked perfectly well, and then contract this disease. It wasn't disfiguring. It wasn't fatal. But it was painful, and you had it for the rest of your life. In a sense, this was a preview of what we're encountering now with AIDS.

A few years ago, Hollywood said it was going to show couples making love in the movies only with condoms. Are efforts like this needed to get your message into popular culture?

There has to be a fundamental shift in our attitude toward condoms. In a casual sexual relation, wearing a condom should be seen as the norm, and not wearing a condom needs to be considered strange or unacceptable. Recently, I was talking to some young people in Geneva who told me there's a real "generational" difference between the sixteen-year-olds and the eighteen-year-olds. The eighteen-year-olds find the condom strange. They think it's not really for them. The sixteen-year-olds don't have the same problem. They see it as a normal part of sex.

The same thing was true for the Pill during the sexual revolution. Depending on what age you were, two years could make a dramatic difference, almost a generational difference, in attitudes. This seems to be happening again today with condoms. I don't know if we'll ever see condom use as the norm, but this would have an enormous effect on reducing transmission of the AIDS virus.

What special meaning is there in the fact that AIDS connects sex and death?

People say AIDS is special because sex and death are linked, but I'm struck more by the sexual dimension of AIDS than by the fact that it causes death. Many diseases cause death. "We can cure those diseases," people say, "but we can't cure AIDS." Tell me, how many diseases can we cure? We cannot cure very many viral diseases, nor can we cure many of the diseases that kill people, like heart disease and forms of cancer. This concept of curing people is relatively new. The strength of physicians used to come from their capacity to accompany. Accompanying a person through an illness towards health or towards death was central to the practice of medicine, but it disappeared when the issue became one of curing.

But isn't the link between sex and death significant?

I know this is always said about AIDS—sex, blood, and death—but it doesn't completely connect with me that way. Yes, it's important that it's a fatal disease and we have no cure. Yes, it's true if we had a cure and a vaccine that people's attitudes toward the disease would probably be substantially different. I just think that the feelings people have about AIDS, the fear, the tension, the anxiety, are much more closely related to sex than death.

Especially male homosexual sex?

Not necessarily. It's a historical accident that male homosexuals were the first group of people in whom the disease was identified. Worldwide, there is more heterosexual than homosexual transmission of AIDS. The fact that AIDS was first described among gay men in the United States identifies it in people's minds as a homosexual disease, and this leads some people to want to insist that if it *is* transmitted heterosexually, then people must be having anal intercourse. That's not true, but let's face it, one thing we've discovered is that anal intercourse is more prevalent than some people would have liked to believe. Again, we're closing the gap between social myth and social reality.

What part does bisexuality play in the transmission of AIDS?

Bisexuality is the most hidden part of sexual behavior in the United States and many other places. Bisexuality is even more difficult for individuals and societies to deal with than homosexuality or heterosexual polypartnerism. Some interesting surveys have suggested that bisexuality may be more common than homosexuality.

Male prostitutes in many parts of the world, when asked, "Who are your clients?" frequently answer, "Married men." It's incredible that we don't know how common it is, but bisexuality is obviously an important part of the picture. Often, people have very simplistic ideas about sexuality. They think it's something imprinted on you at birth or puberty, and then that's it, that's what you are. They don't understand that different pressures create different situations. Among sailors and men in prison, situational homosexual intercourse is a well-known reality.

In Latin America a substantial proportion of men who are infected with HIV say they are bisexual, not homosexual. We don't know yet whether this is a defense against being called homosexual. But we do know that in some societies the penetrating partner in a homosexual couple considers himself to be heterosexual, whereas the penetrated, the passive, partner is seen as bisexual or homosexual. So even the words for our sexuality don't mean the same thing in different cultures.

Luc Montagnier, who discovered the AIDS virus, describes it as a disease of the cities. It is caused, he says, by the increased pace of modern life and the globalization of travel and sexual contacts among people. This leads Montagnier, only half jokingly, to call the Boeing 747 the primary vector for spreading the virus.

AIDS isn't just a problem of the cities, but I agree with Luc that it's associated with stresses on individuals and cultures that have been accentuated in our lifetime, and I do believe there is something unusual that happens to people living in big cities. It's extraordinary to see Africans drawn from rural Zaire to Kinshasa *la belle,* the beautiful city where the lights burn all night. At first people coming from different parts of the country stay together, but then things begin to fall apart. The social fabric turns atomistic. This puts strains on individuals and affects their behavior. The dangerous side of globalism is the loss for individuals of a more rooted group identity. I agree with Luc about the nature of the AIDS epidemic, but I also think there's something deeper going on, something subterranean, like the geologic movement of tectonic plates. I feel in AIDS the shifting of these tectonic plates.

What tectonic plates are moving under the AIDS epidemic?

They have to do with relationships among people. They involve attitudes and belief structures that are deeper than the question of whether or not you can get a ticket on an airplane. It isn't the Boeing 747 and the existence of cities that caused this epidemic. There's something else going on in our world, and these changes have as much to do with recent events in Eastern Europe as they do with our awareness of how diseases cross borders, our understanding of the human rights dimension of disease, and our capacity to see AIDS as a global problem. Something has really begun to click in our attitude toward the world as a whole. We don't have "enemies" the way we used to during the Cold War. At the global level, there are no longer good guys and bad guys. We're all in this new world together.

AIDS may be a global epidemic, but aren't people's perceptions of it regionally fragmented?

It's true that in different part of the world there are different perspectives on the disease. In the United States and Western Europe it's perceived as a problem for people in particular "categories," and a category always means somebody else. It's a disease of "them"—male homosexuals, drug users, prostitutes. This roughly corresponds to the epidemiological reality, but it sometimes doesn't sink in that heterosexual men and women are also infected with AIDS.

In Africa AIDS is seen as a disease affecting men and women alike. Africans understand very well that it's spread by sexual transmission and that people with many sexual partners are at greater risk. Unlike the United States, AIDS is not perceived as a disease affecting "marginal" people outside "the general population." In fact, people with AIDS in Africa have included members of the social and economic elite.

In Asia, AIDS was thought to be a Western disease, and for the early cases this was essentially true. Approximately half of the early AIDS cases in Japan were people with hemophilia who were infected by imported blood products. The same is true in the Middle East, where the first reported AIDS cases occurred in people infected by imported blood. This also happened in India, where one of the first AIDS cases was someone who had gone to the United States for open heart surgery.

The disease was also thought to be Western because of its association with homosexuality. Early reports in the media placed a lot of emphasis on this aspect of the disease. But of course to call homosexuality a Western custom is not accurate. Homosexuality has certainly been a fact in Asia, and bisexuality has been written about in the literature of many Eastern cultures.

While the first AIDS cases in Asia may very well have resulted from direct contact with the West, the epidemic that's occurred since then in places like Thailand is indigenous. Bangkok is estimated to have at least 50,000 intravenous drug users. In December 1987, their infection rate was one percent. Today, it is over forty percent. This is an epidemic fueled by needle exchange among Thai drug users, and, at this point, it has nothing to do directly with the outside world.

Do you hear a lot of scapegoating stories about the origin of AIDS?

AIDS is associated with a sense of culpability, of guilt, of being stained. There's a tendency to blame others for one's affliction. Westerners say it came from Africa. Africans say it came from the West. Asians say it came either from the West or Africa. Haitians say it came from outside Haiti, while others see Haiti as the source of the epidemic. These are probably unresolvable questions. Unfortunately, the search for a scapegoat and the search for the origins of the disease have become associated.

The origin and evolution of viruses is a valid scientific question, but the issue of where AIDS came from is also on the political agenda. I've witnessed some very sad scenes of people claiming to be experts on the subject talking to Congressional committees and the media about "African sexuality." This is as stupid as talking about "American" or "European sexuality." Is there a European sexuality? Not if you know anything about Romania compared to Denmark.

I did a lot of traveling around the world in the early days of setting up the global AIDS program. I'd spend a day or two in one country and then move on to the next. "Where do you think the epidemic started?" I would ask officials in country A.

"It came from the prostitutes in country B," they would say.

Then the next day, when I was in country B, I would ask them, "Where do you think the epidemic started?"

"It came from the prostitutes in country A," they would say. It's incredible how often this scenario gets played out.

The way people view the disease has several levels to it. First, there is the global level. Many people understand that AIDS is a worldwide threat. The idea that this is a world that can be touched in every one of its parts by an infectious disease, especially one spread sexually, is well known by now.

Then at the national and community levels people have a differentiated view. They may say, "It's not a problem for people like me; it's only a problem for *those* people." This means, "It's only a problem for people who perform those acts—injecting acts or sexual acts—that I disapprove of, or that people like me don't engage in."

Then there is a third dynamic, which is one of the reasons again why AIDS has a global dimension. It's hard to say this without sounding rhetorical, but AIDS is so personal that it's universal. Only certain things are universal, and sex is one of them. Maybe sexually transmitted diseases at other times in history were perceived this way, but I doubt it, because people didn't know what was happening in other parts of the world.

What is the history of the epidemic, as we know it today?

I personally believe the AIDS virus has been around a long time. What appear to be cases of AIDS occurred in the late 1950s and early 1960s, and AIDS started to become a worldwide problem in the mid 1970s. During its silent phase, unbeknownst to all of us, the virus was spreading throughout the world, and by the time it was discovered in the United States in 1981, AIDS was already present on at least five continents.

We estimate that worldwide a hundred thousand people were infected in 1980, while today there are over six million. With the possible exception of Albania, which reports no HIV infections, and one or two island nations in the Pacific, the virus has spread to every part of the world. And how confident are we that another undiscovered virus isn't today having its own silent phase of spreading throughout the world?

What could have been done earlier to prevent the spread of AIDS?

No one should ever be satisfied with any delays—and there were delays—but, compared to what could have happened, AIDS was detected relatively early. The AIDS virus became discoverable at about the same time that AIDS became an epidemic. This would not have been the case if the pandemic had occurred in the 1950s, because at that point science didn't know about human retroviruses, and the technology to identify them didn't exist. We didn't have the capacity for isolating the virus, screening blood, preparing tests. Nor did we know much about the human immune system and how viruses infect people and do their damage.

Is the virus transforming itself in order to infect more carriers?

There are only three modes by which AIDS is spread. It's spread through sex, through blood, and from mother to infant. That's it. Limited to these three pathways, the virus shows no evidence of changing its modes of spread. But this still leaves a wide range of opportunities for transmitting the disease.

In North America, Western Europe, Australia, and New Zealand, the virus is following what we call pattern I. In these areas, sexual transmission is predominantly male homosexual, although heterosexual transmission occurs and is increasing. Today, blood transmission is essentially through intravenous drug use. And, finally, there is spread from mother to infant, although relatively little, since the majority of people infected are men.

In pattern II countries—sub-Saharan Africa and parts of the Caribbean—the major mode of transmission is heterosexual. Spread through blood occurs mainly through unscreened blood transfusions and medical injections with unsterilized needles and syringes. Because about half the people infected are women, and they are generally young women of childbearing age, there are large numbers of infected children.

In pattern III countries, which include the Middle East and most of Asia and the Pacific, the virus is relatively recent. It doesn't seem to have come until the early 1980s, and there are relatively fewer cases. AIDS has not yet penetrated into the large populations of Asia, such as India and China, although it threatens to, and this would gravely aggravate the situation.

There is another pattern, which is transitional between pattern I and pattern II. It occurs primarily in Latin America. In the early 1980s, sexual transmission here was predominantly homosexual. Today, in countries like Brazil, sexual transmission is predominantly heterosexual. In other words, it has shifted from pattern I to pattern II. But I should repeat that the modes of spread are not changing. What is changing are questions of who is infecting whom, and what social practices are contributing to the spread of the virus.

Why is transmission of the virus changing in someplace like Brazil?

I'm not sure anybody knows. Bisexual behavior could be acting as a bridge between heterosexuals and homosexuals. In Brazil from 1980 to 1989, 16.3 percent of the AIDS cases occurred in bisexual men. The spread from heterosexual drug users to their partners could be another factor. The same bridges exist in the United States and Europe.

Does this make it likely that the United States will also see a shift to heterosexual transmission?

One way to think of the AIDS pandemic is to see it as a complex interweaving of different, smaller epidemics. For example, the inner city population of New York has levels of infection similar to those in Africa or Haiti. But when you generalize to include the entire population of the United States, those local levels of infection get diluted into a mass of over 200 million people. No matter where you are in the world, you can't think of AIDS as a single mass front of virus moving through a population. It moves at different rates at different times in different groups. But it's always connected with behavior, because behavior—whether sexual behavior or injecting behavior—creates the vulnerability that allows the virus to move.

Is there any part of the world likely to remain untouched by AIDS?

Every society is vulnerable to HIV transmission from many different routes. Let me give you an example. There was recently an outbreak of AIDS in the Soviet city of Elista, which lies between the Black and Caspian seas. This was the last place in the world anybody expected to find an AIDS epidemic. A Soviet citizen had gone to Africa, where he was infected with HIV. After returning to the Soviet Union, he infected his wife, who later delivered a child who was infected perinatally. The wife and child were hospitalized. Then, through syringes not being properly sterilized, the infection was spread in the hospital from infant to infant.

The United Nations estimates there are as many as five million injecting drug users in the world today. So far, only a small proportion of them have been infected with HIV. But the remaining majority of those drug users remain vulnerable to a potentially explosive epidemic of the infection.

A colleague and I have worked out an estimate that there are at least one hundred million new cases of sexually transmitted disease in the world each year. Every one of these cases represents the potential for the sexual spread of AIDS. Another vulnerability to AIDS is created by ignorance. People who don't know how the virus is spread are more vulnerable than those who do. People also become vulnerable for economic reasons. A good deal of the male and female prostitution in the world is associated with economic need. The point I'm trying to make is that vulnerability to HIV infection is not fixed.

Will AIDS ultimately affect the same number of women as men?

Over time, this could be true in a country like Brazil, but it's not necessarily the case in other countries. It's very hard to predict what's going to happen in the next ten years in Western Europe and the United States. It's possible for the epidemic to begin moving into the heterosexual population without being readily recognized, and then when it becomes obvious what's happened, a tremendous amount of time will have been lost.

I had dinner recently with a friend who is a nurse. She told me she is treating a sixteen-year-old homosexual with AIDS who is still sexually active, which means he could be infecting his partners.

We hear a lot of stories about drug users and prostitutes deliberately trying to infect their sexual partners. We don't have any hard data, but the impression I get is that these cases are extremely rare. Whether or not these anecdotes are true, they seem to have taken on a life of their own. A myth has arisen about large numbers of AIDS victims seeking sexual revenge.

People then say, "We need laws to protect ourselves from these maniacs!" as if there were hundreds of thousands of them. This in turn leads to the kind of discrimination and stigmatization that causes a lot of damage. It's like scrapping welfare assistance programs because you've found a couple of flagrant cheaters.

What do you propose to do with HIV infected people who continue to have unprotected sex?

Let's take the extreme situation: a prostitute who is unwilling or incapable of taking the message on board and changing her behavior. Before any punitive measures are applied, we have to make sure that all supportive measures and appropriate legal steps are taken. Sometimes these situations turn out to be false alarms. People say they're doing things that they aren't. But after you've gone through all the steps, you might ultimately end up putting someone in jail for a time, as recently happened in Australia.

But this potentially causes tremendous damage to your prevention and control programs. It sends a shock wave through the community. It scares people away from health and social services, from education, from the support they need. People infected with HIV are infected for life. So what are you going to do? Keep "offenders" quarantined for life? The financial cost, the human cost, the cost to your AIDS prevention and control program could be enormous.

It may be politically satisfying for a community to say, "We took care of that problem!" But this "solution" tends to fall apart when you ask the questions nobody wants to ask. How many infected people, who don't know they're infected, have you scared away from getting tested because they don't want to be put in jail? Will carriers of the virus now be careful *not* to tell anyone they're infected and *not* wear condoms, because this signals they're infected?

How do you get infected by someone with HIV in the first place? You have unprotected sex with them. Well, you have a role in this, and if we can reach you, the presumed client of this prostitute who is not practicing safer sex, then you can put on a condom. You can choose not to have sex with that person. To believe you're taking care of the danger of AIDS by putting prostitutes in jail is to ignore the obvious fact that you can also help people not to have unprotected intercourse with sex workers. We shouldn't be led down the garden path to simplistic "solutions" that actually do more damage than good.

Should AIDS be reportable, like other sexually transmitted diseases?

For statistical purposes there are much better ways to track the epidemic. Anonymous screening is one of them. Knowing the names of people who are infected is only of potential relevance if you can do something for them, and if it doesn't undermine your strategy.

What about notifying the sexual partners of infected people?

We wish the sexually transmitted disease control programs in the world had done a better job of evaluating the effectiveness of partner notification strategies. We don't know how much they really cost, and we don't know how effective they are. They might be a terrible misallocation of resources, or they might be very useful. But we'd prefer that these strategies be based on knowledge rather than faith.

How valid are the AIDS-free certificates demanded for entry into China, India, Saudi Arabia, and other countries?

These certificates don't mean much. Delays lasting several weeks to months occur between infection and the time when a test becomes positive. There are problems with false-positive and false-negative tests, and as soon as this kind of requirement is imposed, a black market in certificates springs up. If an AIDS certificate with a blood test costs X dollars, then a fake certificate will cost X plus a few more dollars—and you don't even have to have blood drawn!

We have plenty of evidence in these testing programs that if people fear their names might be handed over to the government, the number of clients plummets. In one of the Caribbean countries, women coming for prenatal care were tested for AIDS, and if the test was positive they risked being deported. The net result was a dramatic decline in the number of women coming for prenatal care. So not only did you lose an opportunity to educate these women about AIDS, but you also drove them away from programs meant to improve the health of their children.

No program based on coercion or punishment has demonstrated any effectiveness in reducing the rate of spread of HIV infection. We learned the same thing in the past from attempts to lock up prostitutes and other coercive measures for "controlling" sexually transmitted diseases. The message is clear. Punitive approaches to changing people's behavior don't work.

How many people's lives have been ruined by false positive AIDS tests?

Hopefully not many, but the problem is there. A single test stands a reasonable chance of getting a false-positive result. So the protocol to be followed is rather strict. Positive test, repeat positive test, confirmatory test. When looked at carefully, even the confirmatory test is sometimes equivocal. But when all this is done, you still end up with some false-positives, although not many.

How many?

It depends on the background prevalence of infection in your community. I can't give you a number. But it's true that a small number of people who are not infected may test positive. They may be ostracized and discriminated against on the basis of a laboratory abnormality. The AIDS test is as good or better than any other test in laboratory medicine, but it's not perfect. It's crucial for people to understand that AIDS tests are fallible in either direction, whether you test positive or negative. There may be some people who are infected but who, for a long period of time, produce no antibodies.

The idea of an AIDS-free sex certificate is misleading. There's a fair amount of sexual tourism in the world, and sexual tourism is very responsive to threats of danger. In some heavily hit countries word spread that HIV infection was high in eighteen- to twenty-year-old women, but low in twelve- to fourteen-year-old girls, so there was a move toward having sex with young girls, because they were considered less likely to be infected. Then when *they* became infected, the tourists embarked for new pleasure spots. This becomes an increasing problem, as the people most likely to be infected travel to remote parts of the world to have unprotected sexual intercourse, which spreads the infection.

Why do you think the AIDS epidemic will not be over in our lifetime?

People who are infected are infected for life, and infection is continuing to spread. A drug capable of eradicating the virus in a person's body is highly unlikely, which means we'll be dealing with AIDS for the rest of our lives, and people in the next generation will be dealing with AIDS as well.

How many people do you think are infected?

It is estimated that at least six million people are infected. It could be as many as ten million. It's difficult to be precise because the data, obviously, are hard to get. This indicates only the order of magnitude.

What are your projections for the number of AIDS cases by the year 2,000?

We expect a tenfold increase in the number of AIDS cases in the next decade. This means five-and-a-half million new cases of AIDS during the 1990s, compared to the six hundred thousand cases in the 1980s. In my opinion these are conservative estimates. I personally think the global epidemic is continuing to gain momentum. If AIDS really begins to spread through the Indian subcontinent, China, Indonesia, or even into Nigeria, where the rate of infection remains quite low compared to the rest of Africa, then the numbers I'm talking about could be way off. But the message is clear. The decade of the 1990s—whether you take into account HIV infections, people with AIDS, or the social, economic, cultural, and political stresses associated with AIDS—is going to be much more difficult than the last decade.

Could the AIDS virus be hundreds or thousands of years old?

Yes, it could. So then the question becomes, where was the virus hiding? Did it change? Did people's behavior change? The history of infectious diseases tells us that epidemics are more associated with human factors than changes in bacteria or viruses. Virologists have speculated that there might be a small group of people in some remote part of the world who have been infected with AIDS for a long time. They thereby achieved a kind of symbiosis with the virus so that it wasn't so destructive. But then it escaped from this small group to infect the larger population, which is being severely affected by a new infection. Or it could have been an animal virus that jumped into the human population. That, too, is a possibility, but who knows?

Is it likely that AIDS originated in Africa?

Knowledge of the origin of things is rarely useful. Take the origin of the human species, for example. When you look at the polemics on the subject, how have they ever helped anybody deal with an actual problem? This kind of thinking about origins has political consequences. It leads people to search for scapegoats and stigmatize the "other" by saying, "It came from there. They did it. It's their fault."

I thought it was medically important to trace the origin of viral epidemics.

The modern world is built in such a way that certain diseases can be transmitted with great speed. Let's conduct a thought experiment. Pick a country that you think is fairly isolated.

Indonesia.

In 1985, Indonesia had about three quarters of a million visitors. That's a lot of people. Tonga had over ten thousand. Papua New Guinea more than thirty thousand. People are moving around a lot more than they used to.

But think how much worse it could have been if AIDS were spread through droplets, through coughing and sneezing, rather than through sexual contact. I assume there are other bacteria and viruses out there that are occupying small ecological niches in the world, that are adapted to one place and one environment, that have been there and never moved around. But the requirements for their transmission could readily fall into place in the modern world.

The plague came to Europe in the fourteenth century through trade routes. This is a very old story: trade routes, communication, linkages between places, movement of disease. Diseases move with people, and they move with the things that move with people. I assume there will be more epidemics as world culture and economies become more integrated. Therefore, we need a system designed to look for unusual viruses or health events that are linked together in some way. This surveillance system would set off alarm bells, but then the question arises: How late can the alarm sound before the situation is critical?

Did the alarm for AIDS go off too late?

There are six or seven reasons why the world could be a lot *worse* off in dealing with AIDS than it is today. As I mentioned earlier, we believe the general epidemic began in the mid 1970s. The disease was discovered in 1981, and the virus was discovered in 1983. Tests to detect antibodies were available in 1985. Even before the virus was identified, we knew how HIV was spread. So public health measures to prevent transmission were already starting. That's a fairly rapid response.

Now, what would have happened if the capacity to detect human retroviruses had not been developed in the late 1970s? People would have gotten sick without any way of identifying the virus. What if the latency period from infection to disease, instead of being nine years, was eighteen years? That would have meant another decade of spread before detection. What if AIDS, instead of manifesting itself through opportunistic infections that were unusual, things like Kaposi's sarcoma and *Pneumocystis carinii* pneumonia, had manifested itself through bacterial pneumonia or other common problems? That would have delayed detection. What if instead of appearing in a readily identifiable group of male homosexuals in Los Angeles in 1981, the disease had been more widely spread in the population?

It's easy to construct a scenario in which AIDS had another five to ten years of silent and unrestricted spread before the epidemic was discovered. And everything we've been able to do to reduce the severity of this epidemic is associated with our capacity to act quickly. Even as it was, during the early years of this epidemic we were unaware that a virus was spreading. In fact, what's to say that another unknown virus isn't spreading worldwide today?

I thought the World Health Organization was responsible for detecting new diseases.

There's a difference between monitoring existing diseases and setting up a net to catch new ones. Despite the tragedy of AIDS, there's a lesson to be learned from the experience. Denial has been an integral part of this epidemic from the very beginning. Denial at the individual level, the community level, the international level. I'm not saying we handled the epidemic perfectly. Not at all. But shouldn't it be telling us something? Can't we see in it truths that otherwise would have been missed? The thing that's changed is not a new virus, but a new world. Even within Europe, look at all the movement across borders. This new world has specific consequences for disease transmission.

How do you feel when you come face-to-face with someone dying of AIDS?

I had seen AIDS cases before, but I first really came in touch with people with AIDS, their families and the physicians and nurses helping them, when I went to work in Zaire in 1984. The real heroes are the people on the front line, and the front line is where people are caring for people with AIDS or people who are infected with HIV. They're not passively submitting to the disease. They're living with AIDS. They've helped all of us learn that people infected with a virus have choices, and one of those choices is to live with the virus and with the disease, and not just be a patient or victim.

What did you accomplish during your two years in Africa?

Projet SIDA was the first comprehensive AIDS research project in Africa. We clarified some basic issues. How is the virus spread in Africa? There were a lot of rumors and half truths flying around, but the Project did the studies. We looked at prostitutes. We looked at households. We looked at the possibility of AIDS being transmitted by mosquitoes.

Politically it was also important to be developing AIDS data from Africa. This was a trinational project, involving Zaire, the United States, and Belgium, that functioned under the guidance of the Zairean Ministry of Health. We were not foreign expatriates parachuting into Africa and secretly sending out sera to be tested in laboratories in the United States. The Project was a real international collaborative effort of the kind that we need to see more of.

For a doctor accustomed to working with the powerful tools of modern medicine, things like drugs and vaccines, doesn't it sometimes seem ridiculous that all you can offer in the fight against AIDS are condoms and pamphlets?

And education and prevention and support. We can do a lot to help stop discrimination against infected people. That's important. In fact, the discovery during the 1980s that it's essential to protect the rights and dignity of infected people in order to fight this epidemic is a major public health discovery. We've learned that some of the traditional methods of public health, like quarantine and isolation, have never really been effective, except under very unusual circumstances. When people try to apply punitive and coercive approaches to the control of AIDS they don't succeed. The only programs that are working are those based on respect for human rights and dignity.

We can also do a lot to help mobilize resources, raise awareness, increase political and social commitment, and help ensure that laboratory testing is of good quality. Accompanying people who are trying to figure out how to live with a disease that may kill them is also very important. There are lots of things we can do, short of curing the problem. Curing a disease is one thing, but preventing it is perhaps even better.

The Black Death struck everybody indiscriminately, and its victims died fast. Mimi's death from tuberculosis in* La Bohème *seems an artistic, almost poetic, way to perish. Death from AIDS, on the other hand, is prolonged and terrifying.

This may be true, but finally all you may be able to say about AIDS is that this is our world, this is our epidemic. So what if more people died of the Black Death than have currently died of AIDS? Let's hope it stays that way. AIDS is here in our world now. It's us. We're coming to the end of the twentieth century, the end of the millennium, and we're facing a worldwide problem that already has had tremendous impact on modern life.

But while facing the danger in this crisis, we also have to seize the opportunities. The Chinese symbol for crisis is composed of two ideograms, one of which says danger, while the other says opportunity. In every crisis there is danger, and there is opportunity. It's our responsibility not only to face the danger, but also to grasp the opportunities that come with danger—opportunities for strengthening human rights and for promoting public health and global solidarity.

Norman Packard

✦ ✦ ✦

REALITY is messier than physics. Or messier than physics used to be. But thanks to the work of Norman Packard and other members of the Chaos Cabal, physics is now embracing the previously unsolvable mysteries of everyday life. The new scientists of chaos are probing the world to understand how systems evolve, and everywhere they are finding surprises. Packard's dream is to put this element of surprise back into a scientific enterprise that for centuries has been thinking only about laws and order.

The Chaos Cabal started in the late Seventies as a rogue band of physics graduate students at the University of California at Santa Cruz. As a founding member, Packard was an early instigator in this scientific revolution. Using computers salvaged from the physics department basement or cobbled together from scratch, Packard and his colleagues Robert Shaw, Doyne Farmer, and James Crutchfield surprised the scientific community by discovering that many aspects of nature are intrinsically unpredictable. Even simple deterministic systems that seem to have precise rules for going from one moment in time to another, if projected far enough into the future, will develop the loops, folds, spirals, and other telltale signs of order giving way to chaos.

In the process of throwing predictability out the window, the Dynamical Systems Collective, as the Cabal was formally called, took delight in finding chaos almost everywhere they looked—in heart attacks, epileptic seizures, stock market crashes, weather patterns, and even dripping faucets. Instead of being straightjacketed into solvable linear equations, reality was now allowed to run wild in the nonlinear, dynamic patterns it actually prefers to take. And as they watched the everyday world around them dissolve into chaos, chaos itself was discovered to be a realm full of beautiful mathematical patterns and a surprising degree of order.

Mother to this new era of scientific invention was the computer. It alone has the patience to project simple systems far enough into the future to see where chaos begins to wobble their trajectories. And no one is more agile than Packard and friends at picturing chaos and coaxing other creative feats out of computers. Even before the Chaos Cabal came the Eudaemonic Pie, a now-fabled attempt by Packard, Farmer, and friends to beat the game of roulette with toe-operated microcomputers. For five years the Eudaemons risked life and limb in Las Vegas to try out successive configurations of gambling hardware and programs. Knocking over Newton and the casinos were Packard's two early claims to fame.

Born in 1954, the son of a high school mathematics teacher, Packard grew up in Silver City, New Mexico, along with fellow Cabalist Doyne Farmer, whom he met in Explorer Scouts. After attending Reed College and the University of California at Santa Cruz, Packard went on to a NATO fellowship in Paris and several years in Princeton at the Institute for Advanced Study. He has worked as a professor of physics at the University of Illinois' Center for Complex Systems and served as consultant to organizations ranging from the Santa Fe Institute to the Italian government. Married to his former Italian tutor, he spends much of his time in Milan and Turin, where this interview took place.

Today Packard is embarked on various projects that are outgrowths of his work on chaos. One involves the study of evolving systems—things like snow flakes and the stock market; another attempts to model evolution itself. Packard, Farmer, and colleagues have also started a firm called Prediction Company, whose goal is to chart the chaos of financial markets. Allied in these ventures as both tool and subject of his mathematical art are computers—thinking machines that Packard expects quite soon will start evolving on their own.

✦ ✦ ✦

You were involved in the Eudaemonic Pie experiment for many years. Why did you stay with it?

We were convinced it was possible to beat the casinos and felt compelled to realize this conviction. That was perhaps the root of the obsession: proving to ourselves as well as to the world at large that it was a viable program. Our scientific interests are in dynamical systems and prediction, and so there was a kind of coincidence, too, that fed the fires of obsession—although it's not clear whether we learned from that project more about the science of dynamic systems, microcomputer design, programming assembly language, or other things that maybe aren't scientific knowledge.

Will you ever go back to the casinos?

Absolutely not! At least not to the casinos of Las Vegas. But at the moment I'm going to the casinos of Wall Street and Chicago's futures market. Our interest in financial markets has flowered into our starting Prediction Company, cofounded by Doyne Farmer and me. We're teaming up again with some Eudaemonic colleagues to apply complex data analysis to financial markets. We hope to start trading fairly large funds soon. We have a business manager and corporate structure, so cosmetically it's quite different from our gambling forays in the early days of the Eudaemonic Pie.

How will your analyses of the stock market and evolutionary behavior come together?

If the stock market or biological species are changing statistical contexts on you, the learning algorithm, or program, doing your data analysis has to know what kinds of evolutionary changes are possible. And it must respond to these changes. The two realms will meet when computers become sophisticated enough to recapitulate evolutionary changes. For example, in economic systems, the algorithms assume everybody out there is implementing some kind of average behavior. But in fact, people are gradually changing their minds and changing their strategies in evolutionary ways. Right now, the learning algorithms have no way of taking those changes into account. The two approaches will meet when people's gradually changing attitudes and interactions within the economy can be modelled using evolutionary models.

You're saying that the marketplace and people in their economic life display behavior resembling evolutionary processes?

That's right. In fact, people's changing attitudes and behaviors *are* evolutionary processes.

How do you model evolution?

You begin by characterizing the defining properties of evolutionary process. One is an increase in complexity; another is an increase in information processing capability. Still another is the constant ability of the biosphere to generate new possibilities, which then come to have a function or purpose in the biosphere.

Another problem is figuring out if evolutionary change has a sense of direction. Some people call this direction "progress," but this is a loaded word, and I'm not sure I want to say that humans represent progress over bacteria. I'm writing a paper with a section on how evolutionary processes are teleological. Teleology is the study of things that seem to exhibit purposeful behavior. Teleological explanations have fallen out of favor because the fundamental laws of nature allow you to derive the consequences of various actions without dealing with their purpose. But many aspects of evolutionary process cannot be derived from fundamental laws, at least in the same way that you can derive the trajectory of a missile.

What is the biosphere?

By biosphere I mean any collection of interacting living things. *The* biosphere means all living things on earth. Interacting organisms in the biosphere form a complex web, and it's never so simple that one organism exists solely for the purpose of another. Human beings may serve a necessary function for the algae that grows on our pollution or for the viruses that live in us, but that's not the only purpose we serve.

How is chaos theory linked to evolution?

One of the most intriguing aspects of chaos is that very simple, deterministic systems can generate information as the system evolves. You put in simple rules and starting conditions, stand back, and observe all this very complicated stuff coming simply from the trajectory of the chaotic system. This is surprising, and it almost feels like you're getting something for nothing. Chaos is an example of how you produce an ongoing series of complex outcomes from simple beginnings, and this is the kernel of my fascination with evolving systems.

Evolution is even more complex than chaos, because an evolving system is becoming increasingly complex. The fascination for me again lies in this general idea that you can have simple beginnings that produce ever more complex results. This is the link running through both my work in data analysis and evolutionary model building. Since I begin with simple things, I don't have to be too smart to start the process. Then I let the world evolve to show me its complexity.

What did your first models of evolution look like?

Doyne Farmer, Alan Perelson, Stuart Kauffman, and I began by constructing stripped-down models for the origin of life, the immune system, the economy, and the biosphere—all of which are undergoing evolutionary change. It was while working on my simple model biosphere that I developed a way of measuring a system's teleological behavior. My biosphere has little one-celled organisms moving around in a two-dimensional world. The only other thing in this world is food, by which I mean a source of energy. Basically, the organism's genes encode a strategy for moving toward food. I'm trying to get my bugs to learn this task as they evolve. In real terms, you might think of my organisms as bacteria in a petri dish trying to find sugar. When an organism has enough food, it can reproduce, and the genetic makeup of the new organism will be slightly different from that of the parents.

Have you recorded any mutations into variant species?

Species in the real biological world are the product of selective reproduction. My little creatures have no such discrimination. They reproduce indiscriminately. But groups of organisms with similar characteristics do emerge, and I'm now trying to find whether these groups are stable enough to be considered separate species. An advantage of a world within a computer is that it forces you to define concepts like species in the simplest possible way. Ideally, the thrust of this work is to slash away at some of the detail of biology and discover the few elements that capture essential evolutionary behavior.

Is your biosphere similar to Richard Dawkins's?

In my biosphere random genetic changes alter my organisms' survival strategies. I'm not reaching in there and twiddling knobs on the organisms, as Dawkins does with his biomorphs. He accepts or rejects mutations on the basis of aesthetic appeal. My bugs aren't as photogenic as Dawkins's, but organisms in my biosphere live or die solely on the basis of whether a genetic change is functional in the environment. If an organism gets enough food, it will persist.

So in a sense the organisms are learning during their evolution?

Precisely! A population of organisms, with each one shifting its survival strategies during reproduction, is implementing a learning algorithm. Lineages of organisms are learning to survive, but their learning is limited by the fact that their world is self-contained. They are adapting to their environment, which is constantly changing due to the presence of other organisms, but it's very hard to tell just what goal the organisms are learning to achieve. There is nothing outside their world, except for reproduction and the attainment of food, that they are trying to maximize. They're not trying to predict the value of IBM stock, or anything like that. But the next step is to give the algorithm an explicit goal.

What applications might such learning systems have?

They'll help us understand creative processes. The principles underlying evolving learning systems could be used to make thinking machines. Instead of merely organizing knowledge, learning algorithms ask the computer to perform exploratory tasks and discover things it wasn't explicitly programmed to do. Learning algorithms, model biospheres, and other genetic methods allow for the introduction of creative elements into computer programs. The creative potential of thinking machines is what intrigues me.

How do learning algorithms "learn"?

When a computer confronts so many models that it can't go through and try every one of them, it has to have a more general way of analyzing data. In this case, I use what's called a genetic algorithm. It moves around the models using operations that are analogous to point mutations and crossovers in biological reproduction. The computer keeps making new populations of hypotheses that become increasingly fit.

How do computer programs evolve?

Nature can make a mistake during the reproductive process, inaccurately copying a gene. This is called point mutation. Another kind of genetic alteration happens when two parents produce an offspring with a mix of their genetic material. This is called crossover. In my learning algorithm I use genetic operators that are analogous to point mutations and crossover. Once a hypothesis is represented in symbolic form, in strings of symbols, a point mutation takes one of these symbols and changes it to another one. A crossover takes a couple of symbols and replaces them with some subset of variables. The idea is to watch how well your populations perform at each successive generation, and just keep on going.

An organism fit for one kind of world could die out in another.

A good learning algorithm will not only formulate a fit model, it will also keep an eye on the data and let you know when the environment is becoming volatile. There is always the possibility of some catastrophic change altering the environment. How discontinuities happen is not well understood, but the fact that they do happen is unambiguous in both biological and economic evolution, as you can see in a stock market crash. These abrupt changes can come from external shocks to the system, like a meteor hitting the earth and causing mass extinction. But they also seem to result from the complex, internal dynamics of the evolutionary process itself. My ultimate goal is to build a learning algorithm that can cope with evolutionary processes.

What can I look forward to in thinking machines?

The first successes in designing creative algorithms are just coming in. The world champion computer poker program is a learning algorithm. Learning algorithms are being applied to industrial processes, robot control, and speech recognition. These are fairly specific tasks I'm describing. Eventually I think the creativity of machines will be most striking when they start interacting more with people, and when the algorithms controlling the machines are allowed to have more creative flexibility. This isn't too far in the future, and when it starts happening there will likely be a radical change in our perception of computers. They will undergo a transition from being dumb brutes to something much more interesting.

You have said that thinking machines will change the world more profoundly than the industrial revolution.

The profundity of the change makes it difficult to foresee exactly what's going to happen. The change may very well sneak up on us. We're already dealing with machines and interfaces that are fairly intelligent, like the French Minitel system. Gradually these machines and other everyday objects in our homes will get smarter and smarter. Machines will be able to perform many of the tasks that human beings do now, and so humans will be allowed much more freedom.

Will these smart machines include things like voice interfaces?

The creative aspects I'm talking about are not really in the fabrication of a voice, but in what the voice chooses to say. Right now even very simple interactive programs that involve some choices on the part of the computer can produce striking results, like the LIZA program, which is a kind of psychoanalytic game. People interact with these programs quite easily, and they take them seriously.

But in the LIZA program, the machine is only identifying key words and giving predetermined answers.

The point I'm trying to make is that people, if the context is appropriate, are willing to be fooled, and this will aid the transition to the era of thinking machines. We are always trying to animate the things around us, and this tendency will help to animate computers, as well. But, in point of fact, computers *will* become more animate.

What do you mean when you talk about artificial life?

Artificial life is to life what artificial intelligence is to intelligence. Living processes have characteristics that are independent of the material details involved in their implementation. Artificial life is the study of these living processes, which we hope to isolate and implement in other, nonmaterial contexts.

What are these living processes?

You're asking me for a definition of life, and the definition of life is still forthcoming. But if we look at our biosphere, we see that living processes involve different elements interacting with a high level of complexity. These elements process information and make complex decisions. They reproduce and change during reproduction, and these changes are constantly being incorporated into the living process. This last aspect is possibly the most important, because you find things that satisfy the other characteristics without being alive. Calculators and Cray supercomputers have complex structure and do information processing, but they're not alive. Neither of them is able yet to incorporate new information into an ever-growing web of interacting elements. I'm talking about the day when computers can recapitulate evolutionary changes.

What will you do with these new tools?

The new generation of smart algorithms should prove immensely helpful in making policy decisions. When these algorithms can perform as well as I envision, they should become an integral part of the United Nations and every major government. They should make it easier to find social, economic, policy paths to stable, productive situations that now seem to be eluding us all over the world. Debt-ridden national economies struggling to attain some kind of equilibrium by making radical economic changes provide a classic example of hard decision making that involves many, many factors. Ideally, these decisions will be easier to make with the right kind of tools.

Is this the kind of work you do at the Santa Fe Institute?

Yes, developing these tools is one of the reasons why John Reed, the president of Citicorp, funded the Institute. He imagined putting together theoretical economists, natural scientists, and computer scientists to synthesize new perspectives on large-scale global problems. One major thrust that's evolved from his original idea is the development of evolutionary models and learning algorithms to tackle these large-scale social issues and search for nontraumatic paths to equilibria.

What other projects are you working on?

Before looking at the stock market I was thinking about complex spatial patterns. I started growing dendritic crystals to analyze their dynamics. The classic example of a dendritic crystal is a snowflake, but snowflakes are a pain to grow in the laboratory, so we've been using easier crystals, like ammonium bromide. We collect data by looking at the crystals through a microscope connected to a video camera, and then the data is stored and analyzed in a computer.

How did you get from snowflakes to the stock market?

I began generalizing this research out from the analysis of spatial patterns to more general kinds of data. Through a typically Italian series of family connections and friends of friends, I got connected with a group of people analyzing data for *la Regione Lombardia,* the northern Italian state of Lombardy. The state hands out money to seven hundred fifty offices, and each office then uses a different set of procedures in deciding how to spend the money. Meanwhile, back in Milan, the office of management and budget wants to know which of these procedures is most efficient.

And chaos theory can help?

For five years the offices have been filling out forms detailing their decision-making procedures, but no one really knew what to do with the information. After talking to the people involved, I figured out a way to adapt my analysis of complex patterns to this particular problem. The method should be general enough to apply to many complex situations, like the stock market or industrial or other decision-making processes, and right now I'm looking for different sources of data on which to use the tool.

What did you tell the government about how to run northern Italy?

This is hard to say, because when I got the data it was already reduced to a sequence of numbers. My program just looks at the numbers to find patterns. I have a key somewhere that identifies what the numbers mean, but I generally don't pay attention to that. Apparently the director of the office of management and budget is very happy with the results, and the Italian government might even implement the program in other regions of the country.

So how does this apply to the stock market?

I'm developing a learning algorithm that is particularly good at analyzing data from systems that have many, many degrees of freedom, like the stock market. An algorithm is really nothing more than the sequence of commands that tells a computer to do something; it's a computer program. And a learning algorithm is a particular kind of program that's good at searching through a variety of models and finding one that best fits all the current and future data.

How do you define "chaos"?

Chaos is a particular kind of random motion, one that combines both randomness and structure. The system starts somewhere in space and goes along until it falls onto an attractor. With the simplest kind of attractor, a fixed point, the system goes toward a single state and stays there. You see this with a marble in a bowl; it rattles around until it reaches the bottom of the bowl. With a periodic attractor, the system cycles through a sequence of states, like the arm of a metronome left to right and back again in a regular cycle. If you perturb it briefly, it tends to return to its set cycle. The next most complicated kind of attractor—the strange attractor—displays both structure and chaotic motion.

The intrinsic randomness of a chaotic system limits predictability. But the structure of the attractor implies that you can predict part of the time what the system will do. Chaos represents an indeterminate level of predictability between the motion of the planets, which is derivable, and something completely random, like particles in Brownian motion. Chaos presents systems that are random in the long run but have just enough structure so that in the short run you can figure out what they're going to do. The name of the game for my data analysis techniques and learning algorithms is to probe that limit. How far into the future can you predict?

How do you recognize chaos?

Random behavior in a system that is mainly deterministic reveals a particular property of chaos. Two trajectories starting out from nearby states will ultimately end up very far away from each other. Small initial differences are expanded by the way the system evolves. This has been called the butterfly effect, or sensitive dependence on initial conditions, or information generation. Say a leaf runs down a babbling brook. If you drop another leaf in the brook precisely where you dropped the first, it might do the same thing for a little while. But soon it will do something completely different. One reason is because you didn't put the leaf in the brook at exactly the same place as the first. And this slight difference gets magnified into completely different behavior.

Can evolutionary changes be caused by chaos?

They certainly can be. Evolutionary shifts can come from two possible sources: external shocks to the system, and the intrinsic dynamics of the system itself. Even if the environment continues unchanged, all of a sudden something radically new can happen in the behavior of the system, and this radical turn might very well be triggered by some chaotic process.

What examples of chaos exist in our everyday world?

A dripping faucet is Rob Shaw's classic example of a chaotic system. Water running slowly out of a faucet will drip periodically. Water running fast will flow smoothly. But somewhere in between is a regime where the water drips erratically. If you represent it the right way, this erratic dripping displays not only randomness, but also the very definite structure of a strange attractor.

What is the structure you see in chaos?

There are many different kinds of randomness in the world, but by chaos I mean something that in principle needs only a few variables to pin down the behavior of the system. It's actually a very complicated thing, requiring a tremendous amount of information, to describe water dripping from a faucet. Yet with all that complexity in a detailed description of dripping water, the system can still be collapsed into the simple structure that produces the chaos. Other kinds of randomness don't have this collapse into a simple form. A lot of randomness is produced by something more complicated. And I wouldn't call that chaos.

Are heart attacks and epileptic seizures examples of chaos?

Heart attacks used to be considered a common example of chaos. They may in fact be due to transitions from one kind of dynamic behavior to another. But it's not necessarily clear that chaos is always the bad guy. In fact, there's some evidence that fibrillation is a very rapid, highly periodic oscillation, which means it isn't chaotic. Our normal everyday heartbeat, on the other hand, if you look at the timing very carefully, shows considerable irregularity. Another example is in brain waves. It turns out that you have extremely precise oscillations that are not chaotic in the midst of an epileptic seizure. So again chaos may not be the bad guy.

How can something be both deterministic and chaotic?

Historically, this discovery came as a big surprise. Science for many centuries has been preoccupied with a certain kind of powerful technique based on fundamental laws. These laws took the form of equations, and you derived things about the world by solving these equations. Chaos represents an instance where you have simple equations producing behavior that is *not* derivable. Science for centuries had been moving along the track that tacitly assumed strong derivability. Maybe we didn't have all the mathematical tools in hand, but we thought that in principle we could develop them and thereby gain power over reality. Chaos provided the first example where this program saw defeat. It showed us very simple deterministic equations, which one might previously have thought *should* be solvable, but which in fact produced random, underivable motions.

Has the Newtonian universe been overthrown?

Yes, people now accept the fact that there are large domains where strong assumptions about predictability no longer apply. There are still large domains where they do apply. But science moves along pretty rapidly, and these new ideas about chaos got incorporated quite quickly once people understood them.

Does chaos always involve a strange attractor?

Chaos, in the strict sense, is randomness produced from some simple form, and the geometrical form producing the chaos is a strange attractor. A system starts out in some state and then relaxes onto an attractor, just like a metronome. If you kick it a little bit, it gets jostled away from the attractor and then relaxes back down onto this object that lives in state space.

State space?!

A strange attractor is an object in state space just as an ashtray is an object in space. Except that in a state space the coordinates are not like x, y, z coordinates. They depend on context. Let me give you an example. The state space of the stock market is the value of all the stocks, the money supply, foreign exchange rates, the price of treasury bonds, and so on. These are the state space variables and its coordinates. And just as you can describe the geometrical properties of this ashtray, so too can you describe the geometrical properties of a strange attractor in state space.

Is a strange attractor a force like gravity?

A strange attractor is not a force. It doesn't have the same kind of physical reality as gravity. The state-space picture for a dripping faucet shows you a certain kind of structure. The state space data for the stock market may show you the same kind of structure. These structures can be identified as manifestations of a particular kind of strange attractor. The meaning of the variables for each of these state spaces is different, but the attractor is the same. The science of chaos tries to figure out which kinds of attractors are commonly found where.

Why do your pictures of strange attractors look like lopsided butterflies and other peculiar shapes?

This is just what people found when they discovered these strange attractors. One outstanding question in the field is how to classify these shapes. What are all the possible shapes, and which ones are commonly found in nature? We don't have the answers yet, but there are some people very skilled at writing equations of motion that produce strange attractors.

How would you know if an equation is going to reveal a strange attractor?

You don't know for sure. You have to go to the computer and form an orbit that implements the time evolution of your equation. If the orbit starts executing motion on a strange attractor, it will no longer be following a single line on a smooth surface. The surface will be folded, because nearby orbits are diverging from one another—the butterfly effect. For the orbits not to diverge away into infinity, they are forced to fold back onto the attractor, and if you look closely at the folds, you'll see folds within folds within folds. These successive folds are what give chaotic attractors their fractal structure. A fractal is a geometrical object that has infinite nesting of self-similar structure. All fractals are not strange attractors, but all strange attractors are fractals.

Sensitive dependence on initial conditions and rapid divergence of nearby trajectories—is this the basic law of chaos?

A strange attractor is not something that has physical properties that you usually associate with laws of nature. A strange attractor is an abstract geometrical form that lives in the representation of an observed dynamic. The form of the attractor is independent of the things that are implementing the dynamic. So you don't want to regard the attractor as fundamental in the same sense that Newton's laws of mechanics or Einstein's field equations for gravitation are fundamental.

Santa Cruz mathematician Ralph Abraham wrote, "Many people believe that the connection between chaotic attractors of theory and those of experiments is fictitious."

That statement is out of date by now. Ten years ago people were uncertain whether the strange attractors being illustrated by computers had any relevance to physical systems like dripping faucets. By now we have strong evidence connecting these random phenomena in the real world to the abstract, simple models we study.

Are we surrounded by strange attractors without knowing it?

Right. I predict that in a decade there will be children's computer programs for playing with strange attractors.

Would chaos and strange attractors have been discovered without computers?

Experiments that once involved flasks and electric wires are now done in the laboratory of the computer. The computer provides such a crucial jump in the way we do science because it allows us to move into a realm where not all equations have solutions. Nonlinear systems can display their feathers, and all of a sudden we see they have a whole bunch of plumage that linear systems didn't have. The science of chaos is a process of discovery, and in the process these new kinds of chaotic behavior display themselves. The trick is to figure out ways to view them in their full intricacy.

If you don't have solutions to equations, what do you have?

In the realm of chaos, you can create a picture of the statistical context and models that allow you to predict a little way into the future. But for other, more complex, dynamics, even a coherent statistical picture may be too much to ask. Figuring out how to quantify what's happening in a theory of evolving systems is still an open question. I have a few proposals along these lines. But I'm not convinced I have the whole story. One thing I'm working on is a measure of complexity. Another idea is to measure teleological activity in an evolving system. Eventually I want to measure how creative sparks get taken up by a system, which is also a measure of its evolutionary activity.

Where does chaos fit into the business of being a physicist today?

The physics of chaos didn't exist fifteen or twenty years ago, and now it's a substantial part of contemporary scientific activity. It maybe peaked about five years ago, when every issue of *Physical Review Letters* carried a couple of articles on chaos. Now it's down to maybe one per issue. But this just represents the fad of the moment in the physics community.

What comes after the chaos fad?

Most of the exciting new research has moved away from characterizing simple chaotic systems toward more complex dynamics. With a couple more variables and a whole bunch of systems linked together, you can create complex spatial patterns as well as complex motions. I'm interested in an even higher level of complex dynamics—the dynamics of evolving systems. Building a theory of evolutionary processes I see as the ultimate goal of studying complex dynamics.

How do you evaluate the importance of the Chaos Cabal in the early history of chaos theory?

Our biggest contribution was to push the idea of looking at real data and asking whether it came from a strange attractor or some other kind of random motion. We developed some of the first steps for answering these questions and characterizing strange attractors. The development of learning algorithms to press predictability as far as it can be pressed is an outgrowth of those first steps. This is one of the most important parts of chaos theory today.

James Gleick in his book Chaos *describes you guys as Lone Rangers who single-handedly struggled against the forces of ignorance to make your early discoveries.*

His picture bears little resemblance to reality. We were maverick graduate students trying to push off in a new scientific direction. But it's not true we were the good guys struggling against all odds. The physics department at U.C. Santa Cruz allowed us to get our degrees, which in my book indicates a certain level of support. They may have lacked follow-up support, but that's their loss and somebody else's gain. Although we felt like mavericks during our graduate career, we were actually absorbed into the system with remarkable rapidity.

As scientific revolutions go, chaos seems to have been accepted with little opposition from the scientific community.

You're right. The scientific community at large has shown remarkable flexibility in incorporating a new set of ideas into its traditional framework. But there is still uncertainty in the field about what chaos theory is going to produce. Chaos is a ubiquitous aspect of reality, and it is important to understand its mechanisms, but it's not clear that this will allow us to build a better mousetrap. I wouldn't suggest spending billions on strange attractors and chaos theory as it initially emerged, but I might well suggest spending billions trying to understand evolutionary processes.

You used to aspire to be a Renaissance man. . .

I still play the piano and the game of Go and sing choral music, even if I do all of them more or less badly. I've managed to maintain a certain breadth in my intellectual endeavors, and, in fact, all of us in the Santa Cruz group have been extremely lucky in this regard. We're still carrying on the idealistic aspects of the Chaos Cabal's early research into physics, computational theory, biology, statistics, economics, and art. Rob Shaw, for example, did the special effects for a movie, with many of his images produced by chaotic systems.

Aren't you doing something similar with Italian fabric design?

Yes, I have the idea of taking complex patterns generated by chaotic dynamics and looming them into cloth. This would give people like Missoni a new library of images. I have a write-up on this idea of using chaos as a design tool, and now that my Italian is a little better, maybe

I'll hand it around Milan and see if it provides a spark that gets taken up by the design world.

You've said that one task of chaos theory is to reintroduce creativity into the scientific project.

Science has generally not concerned itself with creative processes, because they involve things that are inherently underivable from fundamental law. Many scientists assumed they would eventually tackle these problems. Today we have the hydrogen atom and a few simple molecules. Tomorrow we'll be able to derive the properties of complex molecules, then cells, and then eventually organisms and brains. But this assumption is absolutely false. There are certain points beyond which you can't derive what is going to happen at the next level of complexity. Evolutionary processes epitomize these kinds of blocks—derivability blocks—and creativity in general presents these blocks.

So what does science do with creativity?

From the Greeks to the present, because science got tied up in deriving things from fundamental law, the idea of change as having its own reality got left by the wayside. There was an intellectual bifurcation where the creative aspects of the world became religious questions. This dichotomy broke down with the advent of Darwinian evolution. But science is still having a hard time formulating theories about evolutionary processes. It's extremely difficult to do evolutionary experiments, because it takes so damn long and the systems are so complex. But computers are now giving science a new tool for addressing these creative phenomena.

In what ways are computers themselves becoming creative machines?

In much the same way you're a creative machine or the biosphere is a creative machine. I think qualitatively the phenomena will be very similar. Computers will become creative by being able to search out ways of thinking and solving problems they were not explicitly programmed to do beforehand. The idea that computers will be creative is surprising, because in some sense you do have to program them. Is creativity inconsistent with a computer being programmed? I don't think so. I think it will soon become clear to everybody that it's not inconsistent to have creative process from a mechanical object like a computer, any more than it's impossible to generate randomness from a deterministic system.

Some people are terrified by the prospect of creative machines, but you seem quite pleased by it.

I expect computers will start participating in our evolutionary reality. There is a human tendency to be worried about the survival of ourselves and our species. But in an evolutionary time scale, these things come and go. I can't really get too depressed about the prospect of humanity not being around forever. Evolutionarily, you would expect something else to come along eventually, and it will be interesting to see what does. The advent of these new participants in the evolutionary process might well give us a chance to see what interesting things will arrive next. I think that's exciting rather than depressing.

What if computers replace their masters?

If you feel threatened, you can always pull the plug. Computers can participate in our evolutionary process in many ways without being antagonistic to us. The fact that new elements come into the biosphere doesn't mean everything else has to die. And even if some elements die, is it really a shame we no longer have dinosaurs? Maybe we wouldn't be here if we did. Maybe if we die out then some other even more beautiful species will be created in the future. I have a certain aesthetic attachment to the process of evolution itself, and if one of our creations stopped this evolutionary process, I would find that offensive. But it's more likely to happen with atom bombs than computers.

Mary-Claire King

✦ ✦ ✦

"I'VE learned not to question the motives of bastards. They just do what they do, and you try to stop it," says geneticist Mary-Claire King. Her tool for stopping bastards? Genetics, the old science of families and heredity that today is being revolutionized by decoding the human genome and following its strands of DNA around the globe and even backwards into prehistory. "I've never believed our way of thinking about science is separate from our way of thinking about life. Whether we realize it or not, we are all political animals," she says. King is a model citizen-scientist, with a remarkable knack for turning discoveries at the forefront of her field into tools for the disenfranchised, be they women, people with AIDS, or victims of Latin America's death squads.

Today King divides her time between the School of Public Health and the Department of Molecular and Cell Biology at the University of California at Berkeley, where she heads a lab of twenty-three researchers. Any one of her accomplishments could make another scientist's full-time career. She has isolated the gene responsible for inherited breast cancer, which affects six hundred thousand women in the United States alone. She has identified the gene for inherited deafness. King has also unmasked genetic differences in how people with AIDS react to the virus. This information will be critical for developing therapies and a vaccine.

In pioneering research that hit the cover of *Science* magazine in April 1975, King established that human and chimpanzee genes are ninety-nine percent identical. Her findings were used to calibrate a molecular clock—the rate at which genetic molecules evolve—which revealed that apes and humans diverged about five million years ago, far more recently than people had previously thought. Her other major evolutionary research focuses on the genetics of mitochondrial DNA, the hereditary material that all of us can trace back through our mothers to a common ancestor, the so-called "mitochondrial Eve," who is thought to have lived two hundred thousand years ago in Africa.

King now directs an international drive to map the mitochondrial DNA sequences of diverse populations around the world. Called the Human Genome Diversity Project, this is a twin to the Human Genome Project, which is working to map and sequence one human nuclear genome. King's project will study and attempt to safeguard the world's mitochondrial genomes, especially those ancient populations, like the African pygmies, who face extinction.

King's political engagement began as a graduate student at Berkeley in the Sixties. An antiwar activist, she dropped out of school to work for Ralph Nader, for whom she studied the effects of pesticides on farm workers and sought genetic markers that would indicate human exposure to DNA-damaging chemicals.

The most remarkable of King's political engagements began in 1984. The Abuelas de Plaza de Mayo, the Grandmothers of the Plaza of May, asked her to help retrieve their grandchildren, who had been abducted during Argentina's Dirty War in the Seventies and either sold or given to military families. The parents of the children are now dead, or "disappeared," as they say in Argentina, along with fifteen thousand other people who were tortured and killed during this fascist reign of terror.

The grandmothers needed evidence that would expose false families and prove the grandmothers' relatedness to children who, if born in prison, these women may never have seen before. King developed an array of highly specific genetic markers for proving family relatedness. Using her genetic evidence, the grandmothers have won fifty court cases reuniting them with their grandchildren, and Argentina has established a national genetic data bank for resolving more cases in the future.

Born in Illinois in 1946, King—a great puzzle-solver and lover of mysteries—studied mathematics in college before she realized in graduate school that genetics is the greatest puzzle of all. She sees herself engaged in scientific and political revolutions that are rapidly changing the world. Her Berkeley office occupies the old command post from which she helped organize student protests against the Vietnam War. King is still fighting bastards, and still doing breathtakingly good science.

✦ ✦ ✦

How did you finish your Ph.D. when you were so involved in the antiwar activities of the Sixties?

It was impossible to do science when Ronald Reagan, who was then governor of California, closed the University and sent the National

Guard to throw us out of the buildings. I was in complete despair. I dropped out of school and went to work for Ralph Nader, studying the effects of pesticides on farm workers. After a year I was offered a job with Nader in Washington and was considering taking it when I went to see my friend Allan Wilson, professor of biochemistry and molecular biology at Berkeley. "I can never get my experiments to work," I said. "I'm a complete disaster in the lab." And Allan said, "If everyone whose experiments failed stopped doing science, there wouldn't be any science." So I went to work in his lab.

What research was Wilson doing?

He was using molecules as his analytical tool to study how species evolve. He postulated that humans and chimpanzees diverged about five million years ago. This was much more recent than people who looked only at the fossil evidence had thought. As one of a variety of tools that needed to be brought to bear on the question, Allan asked me to look at the genetic difference between chimpanzees and humans. I kept thinking the project was a disaster because I couldn't find any differences. I would do all these tests looking at migration rates of proteins, and I'd see a difference in one out of a hundred tests. But humans and chimpanzees are really *very* similar. I was in total despair—my usual reaction to anything I tried to do in the lab—but Allan kept saying, "This is great; it shows how similar we really are to chimps!" He turned straw to gold, and I wrote a perfectly reasonable dissertation that landed on the cover of *Science.*

So how do *we differ from chimps?*

In spite of our genetic relatedness, there are clearly important differences between humans and chimpanzees. Allan said there's got to be a reason for this, and the reason lies not in the structural genes, which we've shown to be very similar, but in the timing of when they are expressed and their regulation. This was straightforward and remarkably clever, although we couldn't prove it at the time. Since then there's been a lot of work in this area, and other people have found the mechanisms affecting regulation. Unfortunately, it wasn't me. I'm not a good enough biochemist.

Why was your finding so important for setting the molecular clock?

The idea of using molecular clocks to study evolution had been in Allan's brain since the mid Sixties. While I was doing genetics, other people were using immunological techniques on primates, and genetic techniques on carnivores and plants. We were building a framework with fixed points of reference so we wouldn't have to invoke extraordinary reasons for how the clock could tick differently for different species. This is what some people were claiming to explain away what we had found. But we wanted one clock keeping evolutionary time for all species.

Why is Wilson's work controversial even today?

When Allan died of leukemia in 1991, at age fifty-six, discussion of "mitochondrial Eve" fell to people who created the data but who lack Allan's sophistication about interpreting it. The discussion centers on where and when she lived, not whether she lived, but it will take us longer to convince our colleagues than it would if Allan were still alive.

What are mitochondria?

Mitochondria code for proteins that are responsible for energy production in a cell. Each mammalian cell can hold thousands of mitochondria, which buffer the cell and keep it working at a good clip. It's clearly to a cell's advantage to have large numbers of mitochondria and make lots of these proteins. All mammals have mitochondria, but why they evolved apart from nuclear genes is not clear.

How do you use mitochondria?

When Allan developed the use of mitochondrial sequencing for evolutionary studies in 1985, the idea became central to my own work. It was the base on which we built our technique in Argentina for identifying maternal lineages. Allan persisted for years with the same conceptual theme—trying to understand the processes of evolution using molecules. In the meantime, I became more interested in genetics.

Explain the idea of mitochondrial Eve.

For any two individuals, no matter how different they are, one can always trace their maternal lineages back to a point where their ancestors shared a mitochondrial sequence. Mitochondrial evolution tells us nothing about our nuclear genes or paternal ancestors. All it allows us to do is trace our purely maternal ancestry. If we trace one lineage, one branch of a branching process, you and I and everyone else can tie ourselves back together. So there has to be a common origin for this branching process.

Using the molecular clock—the rate of mutations in mitochondrial DNA—Allan estimated that we all shared a common maternal ancestor sometime between 150,000 and 250,000 years ago, and that ancestor lived in Africa. The reason for this is simple. In Africa one finds much more variation in mitochondrial lineages than anywhere else in the world. The depth of history in Africa with respect to the genetics of this molecule is enormous. All of us can trace our ancestry to molecules that still exist in Africa.

So what's the debate about?

It centers on another question: What is the best tree we can draw to show these evolutionary branches? After Allan's death, his students published a tree, which was the best they had found among the hundreds of thousands of possible trees. But it isn't significantly better than the next best tree. There is so much molecular evidence that the ability to test one tree against another has yet to be perfected. How do you take an enormous number of human sequences, or sequences from other species, and figure out their common ancestor?

Could this ancestor be located somewhere other than Africa?

If you say to yourself, I'm going to construct a tree that shows the common origin outside Africa, you can do it. You can push the data in that direction. You cannot show by statistical testing alone that this tree is inferior to certain African trees. However, none of this bears on the question, Why is there so much more variation in Africa? The mitochondrial sequence used to study evolution is a stretch of DNA that does not code for any genes. Unlike the rest of the mitochondria, it has not been subject to selection, so it's extraordinarily variable. In fact, it's the most variable in the whole human genome. Assuming that this piece of DNA changes at about the same rate everywhere in the world—because there is no selective pressure on it to change faster in one place than another—then it will have changed the most where it's been around the longest. And there's no question where it's changed the most: Africa. All this confusion about statistical testing happened just after Allan died, and, unfortunately, it's muddied his brilliantly simple concept. I have yet to be involved publicly in the debate, but I will be soon, because of the Human Genome Diversity Project.

What's this?

A very big deal. But let me begin with some personal history. The two main influences in my life in genetics were Allan Wilson at Berkeley and Luca Cavalli-Sforza at Stanford. Luca and I have worked together for a dozen years. Luca and Allan were competitors. They were interested in the same problems, but thought about them from different points of view. Allan thought about mitochondrial sequences and constructing evolutionary trees. Luca thinks about human population genetics.

Before Allan got sick, I became obsessed with the idea that he and Luca had to start working together. And they did. They started collaborating on a project to sample the world, the idea being that we need to understand who we are as a species and how we came to be. The best way to do this is to identify ancient populations that have not yet been genetically devastated by invasion or death. We hope to study both their mitochondrial sequences and nuclear genes to try to get a sense of how variation has evolved and genetic migration has occurred.

What populations remain genetically intact?

Some groups of pygmies in central Africa, whom Luca has studied since the early Sixties. There are also populations in Siberia, on the Anderman Islands off the coast of India, the Basques, some Amerindians, and even some Europeans who have lived in the same place for hundreds of years. These are recent populations, but relatively stable ones—not like you and me, modern urban people. Just as anthropologists record cultures, we will compile a genetic record by asking for hair samples or blood samples and decoding the genes. We want to identify genetic diversity in each population and see how this corresponds to diversity in other populations.

What can we learn from studying ancient populations?

We want to learn who we are, what it means to be human, and how we came to be that way. What is the relative importance in human evolution of climate; resistance to pathogens; anatomy; migration; mutation; and genetic drift, which is chance changes in genes? How is evolution influenced by the size of the population and by who marries whom, which dictates which genes are passed on? These are the fundamental forces of human evolution, and the best way to understand how they actually work is to sample people. But not people like you and me who move around at whim. We don't migrate in response to selection or because of deep cultural roots. The best way to evaluate these forces is to identify people who have remained where they are for a very long time. They are the ones on whom evolutionary forces have been acting in a pure way.

What happened when Wilson and Cavalli-Sforza got together?

Allan was already in the hospital when the three of us wrote an editorial for *Genomics,* which publicly launched the project. I remember writing it during the Gulf War. We were concerned that there might not be a world left to sample. One of the groups we wanted to visit, for example, a very isolated population of Iraqi Kurds, has been devastated.

So you agreed on a common course of action?

We wanted to have a sense of human variation in many different places. This led to a tremendous dispute between Allan and Luca, which was great fun to watch. Allan was firing off letters from his hospital room, and Luca was writing back. Everything I've told you is really Luca's point of view. Allan's perspective was different. "The way to understand the forces acting on human evolution," he said, "is not to sample diverse populations, but to put a grid over the entire land mass of the earth and pick a person at every point on the grid." Not you and me, of course, but an indigenous person, someone whose ancestors have been there a long time. "If you select populations in advance then all you will do is confirm what you already know, that there are populations." Allan wanted to make many populations of one size, and Luca wanted to work with fewer, but larger populations. I tried to get the two of them to see they were not really in conflict, but I honestly think they enjoyed being able to argue about it so much that there was no way they were going to come together.

What did you decide to do?

Use both methods, which is what we would have done anyway. The project is going forward. It has taken on a life of its own. Geneticists, anthropologists, and historians all over the world are involved. People are asking, "Why is Basque a unique language? How did Siberians develop their particular anatomical features? Were the Americas settled in waves or streams? What trees are best for studying genetic relatedness? How many people do you have to sample to make a grid? Which populations can tell us the most about human history?" We hope to identify about four hundred populations, which means many hundreds of anthropologists and geneticists will be involved in this project. Our next workshop will confront the question of what it means to do research with indigenous people who are not themselves scientists and who may have only the remotest idea of why we want their blood. What does informed consent mean in these circumstances?

How much is this going to cost?

Five million dollars a year for five years, which means we're going to have to sell this project to a lot of different groups. We want to add genetic evidence from isolated populations to the kinds of cultural evidence that anthropologists have been building up for years, so maybe the foundations that support anthropological work will finance the trips required for genetic sampling. Other foundations may be interested in the technological developments required for field-based medical research. The major difficulty in collecting cells in the field is getting samples back to laboratories quickly enough for them to be put into cell cultures. Individual labs will do the genetic typing, but no one is going to own this knowledge. It wouldn't work if they did. We will all be using the same genetic markers and sharing common information.

Is this related to the Human Genome Project?

Yes. The Human Genome Project is dedicated to mapping and sequencing the human genome—but just one of them. Our new project is meant to understand the variation among *all* human genomes. We can't sequence the entire genome of everybody, but we want to get a handle on variation. The Human Genome Project has supported our workshops, and it was Luca and Allan's idea that we call ourselves the Human Genome Diversity Project.

What do you do with the Human Genome Project?

The Human Genome Project, in Jim Watson's original manifestation, is at the core of everything we do around here. The Human Genome Project provides virtually all the information on which modern gene mapping and positional cloning studies are based. In this lab we're trying to identify the gene responsible for familial breast cancer and ovarian cancer, and another gene responsible for inherited deafness. Both projects involve a technique called positional cloning.

What is positional cloning?

A large number of genes have now been identified for classical genetic diseases, such as cystic fibrosis or sickle cell anemia. These diseases, clearly transmitted in families, are due to the malfunction of single genes living somewhere on a chromosome. The genes for cystic fibrosis, myotonic dystrophy, and neurofibromatosis have been identified by gene mapping, that is, by tracking critical genes down to their position on the chromosome. Once you identify this position, you can sequence that tiny portion of the chromosome and thereby determine exactly which gene is involved. Although it was always theoretically possible, only recently did this approach become feasible. There are three billion nucleotides out there, and you couldn't sequence all of them. However, with the mapping efforts of the genome project, which will ultimately give us an entire human sequence, this information falls into place much more rapidly.

Why are you interested in breast cancer?

My best friend died of cancer when we were both thirteen. She didn't know she was dying of cancer. She just knew she was constantly in horrible pain from what turned out to be a kidney tumor. I was completely devastated by how unfair and inexplicable it was that a person should die at thirteen. So I thought I would do something about it. I didn't learn biology until years later, but the thought was sitting there in the back of my brain the whole time.

Does your being a woman make a difference?

Why do scientists work on the problems they work on? This is related to the question, do men and women do the same kind of science? We think about problems and puzzle them out in the same way. Cognitively we follow the same pathways. But I don't think women and men necessarily find the same problems interesting. Someone once told me that people do science for three reasons—curiosity, altruism, and ambition. I have friends for whom curiosity is all of it. These are the pure scientists. For me the three motives are mixed in equal proportions. I like problems that give useful answers, and breast cancer is unquestionably such a problem. I was insane to think I could tackle it. But I stayed with it because I am a very stubborn person.

Are you ambitious?

Yes. I want the importance of what I do to be recognized. Do you remember what Robert Oppenheimer said when President Kennedy awarded him the Fermi medal, after he had been smeared all those years by McCarthy? "I tried to pretend that what happened to me didn't matter," said Oppenheimer. "But it does matter that your country cares about what you do." Money is not the issue. It's not that kind of ambition. If you want money, you don't do what we do.

Your original mathematical project with breast cancer involved 1,579 women. What were you trying to determine?

Whether a subset of breast cancer is inherited, and if so, from what gene. Inherited breast cancer accounts for about five percent of the disease. It is transmitted by a dominant gene through mothers and fathers, although fathers are not affected.

One woman out of two hundred will get breast cancer because she has inherited susceptibility to it. That's 600,000 women in America alone. It is important for these women to know they have an inherited genetic disease. It is even more important to the other ninety-five percent of breast cancer patients, because if we can identify the gene inherited in altered form and it turns out to be the same gene that's vulnerable to environmentally induced cancer, this will be critical for diagnosis and treatment of all women facing breast cancer. This research matters, both for what it tells us about an inherited disease, and because the families with inherited disease are models for the rest of us.

How did you find this gene?

It's been clear as far back as the Romans that some families have very high rates of breast cancer. In these unusual families, breast cancer is inherited in the same way as any other genetic disease. This, of course, is not typical. Most breast cancer is acquired, just like most lung cancer is acquired, but there are exceptions. I decided in 1975, when I was a postdoc learning about cancer epidemiology, that I would try to identify the genes responsible for inherited breast cancer. This was a dumb idea, because no work was being done at the DNA level back then. It did ultimately work, but only after the DNA revolution in genetics.

What is the gene responsible for breast cancer?

We have not yet identified this gene, but we know where it lives on the chromosome down to a million base pairs, which is a tiny region of the human genome. We identified families with inherited breast cancer. Then we identified genetic markers situated on all the different chromosomes and determined which markers are inherited with breast cancer, family by family. This is a very systematic approach, but when I first undertook it, it was not systematic at all, because I had to develop the markers as I went along.

Was this research aided by the Human Genome Project?

It took fifteen years from the time I began trying to identify genes responsible for inherited breast cancer until I knew the approximate locale of the gene in question. In 1991, when we decided to isolate the genes responsible for inherited deafness, the same process took two months. That's the difference between technology in the late Seventies and technology today. The next step, which we are doing in the lab right now, is to clone the gene for inherited breast cancer and use it to develop an early diagnostic technique.

I thought you were a geneticist, not a molecular biologist.

I had to shift to molecular biology despite the fact I was never very good at it. You can't do genetics if you don't understand molecular biology.

What is the AIDS project you're working on?

Some people infected with HIV progress rapidly to full-blown AIDS and die, while others progress more slowly. They live with the disease for years. There's tremendous variation, some of which may be genetic. Three years ago we were asked to look at differences in the host genotype. Could variation in immune response genes make some people more resistant to the virus? We have now identified some of these genes.

What do you do with this knowledge?

It can be used either for vaccine development or immunological therapies. Identifying the genes that make people susceptible to disease allows you to understand the interaction between the virus and the human antigen produced by these genes. In the case of AIDS, we look at how the protein folds differently and how the virus attaches to it. This allows us to study the interaction between the virus and the antigen, because we know that some of these attachments work better than others. This is helpful if you are trying to develop a drug that prevents the attachment in the first place, a vaccine that protects human antigens against the virus, or an innocuous molecule that will mimic the virus. This information will also be important when public health services test vaccines or therapies. They need to know if other variables besides their treatments are influencing survival rates. If all the people receiving an experimental treatment are doing better, is this because of the treatment or a genetic influence?

Why is this research moving so quickly?

There's a revolution taking place in our ability to study genes at the level of DNA. We can now take a sample of DNA from any organism, whether a plant or a person, and introduce that DNA into a carrier organism, like yeast, and manipulate it—cut it up, sequence it, see how it's expressed. The transition from basic research to human applications was so fast because DNA is DNA is DNA. You can manipulate it the same way no matter where it comes from. This repertoire of technology for understanding what DNA is and how it works, how it's replicated and expressed has developed enormously in the last ten years. The genetics revolution has democratized molecular biology. It has made it accessible to mathematicians, and you can't get more accessible than that!

How does human DNA grow in yeast?

The fundamental reality of evolution is that we all share a common origin, which means humans are related to yeast. The branching of taxa has occurred in a chronological way that we can reconstruct, and we share many sequences in common with yeast. This is both a relief, because we can work with our DNA inside a yeast cell, and a great frustration, because when we put a piece of human DNA inside a yeast cell, we sometimes find ourselves inadvertently studying the yeast. Obviously, our sequences differ from those in yeast. But they are identifiably the same genes. We have housekeeping genes and so does yeast, and they keep house for us in the same way they do for yeast.

Tell me about your early interest in science.

I was not particularly interested in science as a child. I grew up in a very traditional home in Illinois and figured I would go to school and get married. My father was sixty when I was born. He was a person largely from the nineteenth century. He was responsible for labor relations at Standard Oil. Both my parents encouraged my interest in math, but neither of them was remotely academic. This was not an intellectual household.

What did your mother do?

She worked for the War Labor Board, and I suspect was very good at it, but then she stopped working after the war. Women didn't work in the postwar era, if they could afford not to.

Why were you interested in mathematics?

I like solving puzzles. Mathematics is good for puzzle solving, and genetics is the very best puzzle solving there can possibly be. But I wouldn't have known what the word "genetics" meant as a child. Since my early schooling came before the Sputnik era, I was never really exposed to modern science. My brother, who is three years younger, hit that wave, but not me.

Was it unusual for a woman to study math?

Girls didn't major in math in those days, and I have to say I was a below-average math major at Carleton College in Minnesota, where I did my undergraduate work. The guys there who majored in math were very good, and most have gone on to become mathematicians. I knew there was no way I could be a theoretical mathematician, but I liked applied math, so I came to Berkeley in 1966 to get a Ph.D. in statistics. I was nineteen. I had never been to Berkeley, but it struck me as an interesting place, and I thought it was time to get away and do something different.

When did you get interested in genetics?

I was converted by a course I took from Kurt Stern, a German-Jewish refugee who at the time was the world's leading geneticist. He was a phenomenally good teacher. He explained the most mind-boggling concepts in such simple ways that you not only understood them, but you also understood the thought processes that went into solving them. It's the kind of lecturing I wish I could do.

Do you read detective stories?

I'm hooked on detective stories. I love them. That's what I do for fun. I have a stack by my bed at home. I devour detective stories. I've read all the Sherlock Holmeses many times, and my daughter is also addicted to Sherlock Holmes. I also like the V.I. Warshawski detective stories, Agatha Christie's Miss Marple stories, and Antonia Fraser. I guess I'm prone to women writers. But I think my favorite is P.D. James.

Is this required reading for a geneticist?

You have to like solving puzzles. Everybody in this lab likes detective stories. There's a shelf in the other room full of detective stories, so when you finish the heap by your bed at home, you bring them in and swap them. One of my favorite detective stories is *The Name of the Rose,* which Luca gave me. It's one of his favorites also. It combines history with a great detective story based on poisoning. I would love to write detective stories some day. When I'm an old person, I think that's what I'll do. I'll write detective stories with genetics as the key.

How do you use genetic information to solve crimes?

Many forensics cases involve rape or murder. This means you have to be able to identify a potential assailant, as opposed to his close male relatives. Mitochondrial DNA, because it's purely maternally transmitted, would never permit you to do that. The kind of sequencing we do is perfect for determining that two people are maternally related. It's ideal for human rights cases. It's very good if you need to identify remains and say whether a murder victim is indeed the child of this woman. But for rapes and murders where several suspects may be related to each other, it won't serve.

The approach of choice in forensics has been nuclear genes, which is essentially a modern, infinitely more precise version of blood typing. Blood typing has been around since the beginning of the century. If blood found at the scene of the crime and the blood of a suspect are both type AB, this tells you something, because only five percent of the population has blood type AB. The methods used now in molecular genetics are exactly the same, except that you say, "The blood found at the scene of the crime has this specific genotype, and only one in 100,000 people has this genotype." If the suspect has it, this is very strong evidence. Adam Dalgliesh in P.D. James's books is quite sophisticated about the old-fashioned, classical systems, but as for the modern techniques—VNTR, PCR, and mitochondrial DNA—he needs to be brought up to speed.

Can you explain these techniques?

VNTR, which stands for variable numbers in tandem repeats, is a particular type of genetic marker. These repeats are extremely useful in gene mapping and human evolution studies, because they have nothing to do with the product of a gene. They are interstices between genes and consequently extremely variable. They are easy to type, and they allow you to tell one person from another quite quickly, which also makes them good for forensics.

These evolutionary studies are done by polymerase chain reaction, or PCR. This is the technique we exploited so effectively in Allan Wilson's laboratory. PCR allows you to amplify a sequence from an arbitrarily small amount of material, which makes it extraordinarily useful for a wide variety of applications. We use it all the time in our human rights work in Argentina. You can identify sequences from samples collected forty years ago. It will be of great use in the Human Genome Diversity Project. It's important for determining whether babies are infected with HIV, because all you need is a tiny amount of blood to spot the virus. It's got hundreds of thousands of applications, and it's very simple to do.

Does the technique have any problems?

The difficulty with any PCR-based system is that if you have a mixed sample, PCR amplifies everything, contaminating sequences as well as evidence. Forensics doesn't deal with pure specimens taken from somebody's arm. This is blood lying around at the scene of a crime. But these techniques are coming along, and PCR is enormously powerful, because you can start with a very small amount of material.

What are genetic fingerprints?

They are a multiple banding pattern characteristic of each individual. In principle they should be splendid for forensics work, but, because the bands vary in intensity, you are never sure whether you have a recognizable pattern. Alex Jeffreys, a biochemist at the University of Leicester, recently developed a much more sophisticated way of making fingerprints that should be extremely useful.

Joel Cohen at Rockefeller University frequently testifies in court against genetic fingerprinting. What do you think of his arguments?

I just finished a stint on the National Academy of Science's committee on DNA and forensics. The other population geneticist on the committee was Eric Lander, who also spends a lot of time testifying about DNA. Eric and I tried to set down guidelines for doing the mathematics of DNA identification. We want the mathematical principles to be so conservative that any bias will be in favor of the defendant. If the evidence sample is blood group O and the suspect is blood group O, this tells you very little, because half the population is blood group O. If the evidence sample has a specific genotype that is found in one out of one hundred thousand people in the population and the suspect has the same genotype, this tells you much more, but what does it really mean statistically? How common is this genotype in the population? What *is* the population? Is it the entire human species? I and a lot of other population geneticists have spent years trying to answer this question, and there is no mathematically rigorous way to do it, because we don't have the entire human species sampled.

How will you settle the question?

Eric and I established a set of guidelines, a recipe, for how to calculate the expected frequency of an arbitrarily determined genotype in a population. If the best estimate of the frequency of an allele were one percent, but it was found in some population at a frequency of ten percent, we assumed ten percent was the general frequency. With every assumption we took the most conservative estimate, so this estimate could be applied to any suspect anywhere, regardless of race or ancestry. Some mathematical types don't like this. Eric and I don't like it, in the sense that it is not as precise as it could be, but we and the panel decided to be prudent rather than precise.

How do you make these calculations?

We take blood samples from all the populations around, but there is always a chance we could overlook the population containing our suspect. Let me give you an example. Suppose we test four genes. We determine the genotype in each of four loci on a blood sample from the victim. Each gene has two alleles. We do the same thing for the defendant and everything matches. Knowing how common each allele is at each of the four loci is critical. One way to do this is to determine from which population the defendant comes and make an estimate based on that population. Originally, that's what everyone, including me, used to do.

But in America, where most of us are from mixed populations, knowing the frequency of every allele is impossible. So now we use a ceiling principle. Before a case goes to trial, one consults a data bank of multiple populations, which are very different from each other. The four loci will have been typed in each of these isolated populations. Since these populations are selected to be as different from each other as they can be, we hope to bracket in a broad way the likelihood of finding each gene type.

Then what do you do?

Suppose allele A of gene 1 is found in Basques at a frequency of one percent, in Lapps at three percent, and in Mexicans of Mayan ancestry at ten percent. We will assign a frequency of ten percent to this allele, regardless of the ancestry of future defendants. Now, it doesn't matter what population the suspect comes from, since no people in the world have higher frequencies than the ones we're using. We don't care if the defendant is white, black, Hispanic, American Indian, or whatever, since the whole world is in the calculation.

But the whole world has yet to be sampled.

This will be a dynamic process. One can also say, "We will stop after an arbitrary number of populations." There is a point after which it becomes highly unlikely that some population will hop in with a frequency of twenty percent, given that everyone sampled up to that point has been under ten percent. We can determine that point on statistical grounds, but we're not there yet. The fact of the matter is that we're dealing in court with real people from messy ancestries—like you and me. This calls for good, practical ideas. We'll see in the next few months how many of our colleagues agree.

Are people being convicted on genetic evidence?

Genetic evidence is used at many steps along the way. Most of my consulting involves helping both defense and prosecution sort out what the genetic evidence actually means. Consequently, many cases never go to trial. In at least a third, this evidence exonerates people. Increasingly, every day, genetics goes to court.

But often with hasty interpretations. . .

Not by me, and after the release of our report, we should be moving towards a consensus. What's been happening up to now is that individual population geneticists would do different things when they were asked to interpret the evidence. Lots of people, myself included, got involved in criticizing those interpretations.

You've said there is a "sleaze factor" sometimes associated with genetic research.

I made that comment when the cystic fibrosis gene was mapped with certainty to chromosome 7. At one private company, an officer, a non-scientist, said, "We own chromosome 7." No one would be that ignorant today. The guy thought they could keep proprietary rights on all the interesting probes on a chromosome, which is total nonsense. That you can't do this is even clearer today with the Genome Project. It has enormously facilitated a rapid sharing of information about where markers are located. This includes technologies for typing and sequencing genes and information about physical agents, such as probes.

Are you in the middle of a scientific revolution?

We are at a unique moment in science where it is in everyone's self-interest to share information. There is so much to be done and it is moving so quickly that you cannot possibly do it all yourself. The only way to move ahead fast is to put your information into the general pool and move on to the next step. You'll get to that step faster by sharing information than you would working on your own. So right now even the most competitive people are cooperating.

This doesn't mean everyone shares everything, but it does mean that basic approaches, physical reagents, and technological tricks are shared, and shared very quickly. By the time something is published, it's very old news in this business. The other day I was extolling to a colleague the enormous pleasure of working in genetics at this moment when people are competing to cooperate, and he said, "Enjoy it while it lasts, because once the Genome Project has succeeded, people will revert to their competitive ways." I suspect he's right, but for the moment, we're falling over ourselves to cooperate.

On the flight out, I sat next to a scientist from Genentech who is genetically engineering a vaccine against AIDS. Would this information be shared with other scientists?

Genentech wouldn't give you the sequence of a critical gene unless you were collaborating with them. But I bet I could call them up and get it, because if I show its importance to breast cancer, this would be important news to them. There's no way *they* are going to find that out themselves. They don't have all the families and all the math. This sharing of information out of self-interest happens thousands of times a day all around the country.

How did you get involved with looking for kidnapped children in Argentina?

I first visited Argentina on September 11, 1973, the day of the Chilean coup. My husband and I—I was married at the time [to ecologist Robert Colwel]—were living in Chile teaching at the University of Santiago. I was returning from what I thought would be a quick trip to the States when the coup happened. The military closed the airport, and we landed in Buenos Aires. I was there two weeks before I could get back into Chile, all the time not knowing whether Rob or my students were dead or alive. I found Argentina overwhelming. Perón was in the midst of his last electoral campaign. Already there were battles being waged between the far right and left. I thought, I will never in a thousand years understand this country. It is a most remarkable and strange place. Argentina has since become my second home, and how I ever found it intimidating is beyond me! I was twenty-seven years old and could barely speak a word of Spanish when I got caught in the middle of a revolution.

How did Argentina start disappearing its citizens?

When Perón died, his wife, Isabel, ruled briefly until the military threw her out and imposed an explicitly fascist dictatorship. Both its politics and cultural roots were those of the Italian and German fascists who had migrated to Argentina after the war. Their sons ran the military and took charge after the coup, and in 1975 a civil war started in earnest. The military kidnapped vast numbers of people to terrorize the population. In the process, they garnered pregnant women and women with babies. Children old enough to report what they had seen were typically killed. Pregnant women were kept alive and tortured until they gave birth. The babies were sold or handed out for adoption among the military, and then the mothers were killed.

How do we know this history?

By 1977, a number of human rights groups had formed in Argentina, mostly made up of families of people who had disappeared. One group consisted of grandmothers whose daughters and sons had been killed. Their grandchildren were either born in captivity or kidnapped when they were babes in arms and then sold within the military subculture. Some were probably also sold abroad. This group of grandmothers, the Abuelas de Plaza de Mayo, kept vigil every Thursday in the Plaza de Mayo, the square across from the military government's offices. In addition to their protests and marches and the scarves they wore printed with the pictures of their children, the grandmothers also collected enormous quantities of information.

Typically, they would get reports of a young woman in prison who had been tortured during the course of her pregnancy, until she was ready to give birth. Since no one in these prisons knew anything about obstetrics, a midwife or an obstetrician would be kidnapped off the streets, blindfolded, taken down to the detention center, and told to deliver the child. They were instructed not to speak to the mother, but inevitably they did. They would find out her name, deliver the child, and then see the mother taken away again. The doctor or midwife, after being dumped back in town, would report what they had learned to the grandmothers.

What happened to the mothers and their children?

The mothers were sometimes killed outright. Sometimes they were returned briefly to their cells, where they could tell other women whether they had delivered a boy or a girl. And occasionally someone would be released. At the time of the World Cup finals, athletes refused to play in Argentina unless political prisoners were released, or at least brought to trial. The military went through the cells and picked a few people at random, saying they had been declared innocent and everyone else had been declared guilty. I know this from a woman who was released at the time, someone my age, although she looks about sixty, who is now a lawyer for the grandmothers. Like other women who had been in detention, she kept lists in her head of all the babies who had been born. Janitors turned out to be another good source of information, because the military didn't clean their own torture centers.

How did the grandmothers follow these leads?

When children appeared abruptly in the families of women whom everybody in the neighborhood knew had never been pregnant, the grandmothers would be called anonymously. When these children started kindergarten, the schools had to be presented with birth certificates, and even I can tell when these are forged. They usually claim the child was born at home with no medical assistance. They either have no signature or a signature from a military physician who had nothing to do with obstetrics. By 1983, the time of the Malvinas-Falklands war, the grandmothers had information on one hundred forty-four cases. Now we're up to two hundred seventeen cases. These are cases in which there is evidence that the child was seen alive after birth, left a detention center alive, or was kidnapped alive off the street. We now know that some of the original one hundred forty-four children were killed. But most were not.

How did the grandmothers start recovering these children?

After the Malvinas war, the grandmothers realized the military was on its way out. They knew they were going to be able to bring these cases to court. In many instances, they had hypotheses about the identity of the children, particularly the ones who were kidnapped as infants. Some of the children had been kept by guards at the detention centers where their parents were tortured and killed. But the grandmothers knew they would have to prove their cases. They knew what everybody knows about paternity testing. You can prove someone is *not* the father of a child. But it wasn't enough merely to prove that these imposed parents, as they call them, were not the real parents. The grandmothers wanted to prove who the real parents were.

So in 1983, even before the end of military rule, two of the grandmothers came to Washington, D.C., and met with the committee on scientific responsibility at the American Association for the Advancement of Science. They said, "Here's the situation we confront, send us a geneticist." The AAAS got in touch with Luca Cavalli-Sforza and asked him, "Does this make any sense? Is it possible to prove a child's relatedness to its grandparents?"

Luca did the statistics and said, "Yes, this is a perfectly reasonable hypothesis. Thanks to specific genetic markers—such as human leukocyte antigens (HLA) and variations in DNA sequences—grandpaternity, like paternity, can be determined with a high degree of certainty. In reality, though, there's the problem of sampling all the people involved and getting them genetically typed. This is a different kind of question, and the only way to answer it is for somebody to go to Argentina and find out. I don't have the energy to get involved with Spanish-speaking grandmothers my own age," he said, "but I know just the person you should call. She's lived in Latin America. She's the age of the daughters of these women, and her own daughter is exactly the age of their granddaughters." So they asked me if I would go, and of course I said yes.

When was this?

I went for the first time in June 1984. By that time Raúl Alfonsín was president. He had appointed a commission to investigate all the disappearances, of men and women as well as children, most of whom were murdered. The commission met in a cultural center that was full to overflowing with people sitting quietly waiting to give testimony, thousands of people waiting to talk about relatives and friends who had disappeared. The commission ultimately published a book called *Nunca Mas,* which means *Never Again.* It recorded ten thousand disappearances, but the numbers have gone up since then. I estimate the total number of disappearances at about fifteen thousand. The number of disappeared children we are seeking is two hundred seventeen. The actual number disappeared is probably between three and four hundred.

What happened when you began working with the grandmothers?

I went to Argentina with Clyde Snow and other forensics experts. Their job was to work with the commission identifying remains, so that cases against the murderers could be brought to court. They ultimately trained a remarkable group of Argentinean forensic anthropologists, who at the time were just kids in college. Now they do all the work in Argentina identifying remains. They use classical methods, but also some newer techniques we've been teaching them, like DNA sequencing from teeth. This forensic team has trained other teams in the Philippines, Chile, Guatemala, El Salvador, and wherever else this kind of work is needed.

While they started that training, I worked with the grandmothers trying to identify living children and reunite them with their relatives. We were fortunate enough to find a lab in a Buenos Aires hospital that could do HLA typing. The method works like this. Blood samples are taken from people who might be related. Their cells are tested for matching genetic markers. The most reliable matches for proving relatedness use HLA proteins, which are found in white blood cells. There are thousands of individual HLA combinations in any one population. These proteins distinguish "self" from "other," which makes them important for matching organ transplant recipients, and also for matching grandparents with their lost children.

Did your test stand up in court?

The HLA test we developed was first used in the case of an eight-year-old girl named Paula Eva Logares, who was living with a former police chief and his Uruguayan girlfriend. They claimed in court that Paula was their biological daughter. The grandmothers said they were lying. They said she was kidnapped with her parents when she was twenty-three months old, and the parents were never seen again. After the election of President Alfonsín, when this kind of case could be brought to court, we proved with 99.9 percent certainty, on the basis of HLA testing and blood groups, that Paula was a descendent of the three living grandparents who claimed her. When she went back to her grandparents' house, which she hadn't seen since she was two years old, she walked straight to the room where she had slept as a baby and asked for her doll.

How else did you use genetic evidence?

The first cases were relatively easy, because we either had all four grandparents still alive, or we could reconstruct their genotypes from their surviving children. We got very good at this, and consequently things happened. Many more families came forward. The Argentinean parliament passed a law establishing a voluntary national genetic data bank, so that anyone who had lost a child could have a blood sample taken. We would construct a pedigree, and as children came to light, test them against the families in our genetic data bank and look for a match.

With hundreds of families to test against every child who came forward, we ran into the possibility of getting some matches by chance—false positives. They would be true matches, but they wouldn't reflect biological relationship. So we needed a better test. We also began to learn of cases in which the grandparents were no longer alive. We might have no more to go on than the mother's mother or a maternal aunt—just one branch of the family. That's when we turned to mitochondrial sequencing, which has proved to be a highly specific, invaluable tool for reuniting the grandmothers with their grandchildren.

How many families have been reunited?

The number of cases resolved genetically is about fifty. We have found another twelve children who were kidnapped, but we have yet to identify their families. This leaves one hundred fifty children still to be found.

How much time do you devote to human rights work?

At first I was flying to Argentina several times a year. Now I go about once a year. The grandmothers come here more often. Through friends in New York, I think I know the travel schedule of every human rights-oriented Argentinean who visits North America. So any time that someone comes or goes, they can carry DNA samples or reagents or other material.

This photograph on your desk is your daughter?

Yes. It was taken when she thought she might become a ballerina. She was very good, but she had to decide when she was fourteen whether she wanted to go to school or dance, and she decided to go to school. She's studying to be a constitutional lawyer, not a scientist. She correctly perceives that the limitations to human rights are not in science, but in having a constitution and making sure it's applied. So she wants to do human rights law. But the girls her age I know in Argentina tell me they're going to become geneticists. It all comes out in the wash.

Acknowledgements

✦ ✦ ✦

My greatest debt is to the scientists who allowed me to pester them into granting these interviews and who then took further pains to edit them. Talking about oneself can leave a strange taste in the mouth. I assured my subjects this was socially redemptive work, but I truly appreciate their leap of faith.

Preparing for these interviews reminded me of my doctoral exams, except that in this case the experience was repeated eleven times on subjects about which I was less than expert. I could not have completed this assignment without the invaluable research assistance of Sharon Britton, Linda Brown, Julia Dickinson, Susan Juliano, Lynn Mayo, Kristin Strohmeyer, and Joan Wolek.

For their diligence in transcribing these interviews, especially those portions recorded in piano bars, my great thanks go to David Dupont, Sharon Gormley, and Susan Kovacs.

For financing the travels entailed in collecting these pieces or venturing to publish them in abbreviated form, I am indebted to *Omni, Rolling Stone,* and *Harvard Magazine* editors Patrice Adcroft, Keith Ferrell, Craig Lambert, Susan Murcko, Kathleen Stein, Dick Teresi, Robert Wallace, and Gurney Williams. I wish to single out Kathleen Stein for special praise. She invariably steered me in the right direction and abetted this scheme—even in its dark moments.

I am grateful for the companionship on these intellectual journeys of my wife, Roberta Krueger, our daughter, Maude Bass-Krueger, and other adventurous souls, among them Audrey Bass, Josh Fischman, Richard Harris, Geraldine and Robert Krueger, and Jamie Shreeve. Jack Repcheck, and Lynne Reed did an exemplary job of getting the work to press.

This book is dedicated to my father, Nathan Bass, the scientist in the family. He travelled the world looking for minerals and metals, and I have had the privilege of following in his footsteps looking for ideas. His omnivorous interests, keen mind, and skepticism informed this project from the start. He relished the task and kept me supplied with information on the subjects at hand. His understanding of my attempts to hybridize art and science and his support of a literary career whose direction has often been nonlinear are truly appreciated.

Suggestions for Further Reading

✦ ✦ ✦

SARAH HRDY

Sarah Blaffer. *The Black-man of Zinacantan: A Central American Legend.* Austin: University of Texas Press, 1972.

Glenn Hausfater and Sarah Blaffer Hrdy, eds. *Infanticide: Comparative and Evolutionary Perspectives.* New York: Aldine, 1984.

Sarah Blaffer Hrdy. *The Langurs of Abu: Female and Male Strategies of Reproduction.* Cambridge: Harvard University Press, 1977.

Sarah Blaffer Hrdy. *The Woman That Never Evolved.* Cambridge: Harvard University Press, 1981.

LUC MONTAGNIER

Robert C. Gallo and Luc Montagnier. "AIDS in 1988." *Scientific American* (October 1988), 40–48.

Marie-Lise Gougeon and Luc Montagnier. "Apoptosis in AIDS." *Science* (special issue, "AIDS: The Unanswered Questions"), 260, 5112 (1993), 1269–1270.

Luc Montagnier. *Vaincre le SIDA: Entretiens avec Pierre Bourget.* Paris: Editions Cana, 1986.

Luc Montagnier. "Origin and Variations of Immune Deficiency Viruses." In Jean-Pierre Allain, Robert Gallo, and Luc Montagnier, eds. *Human Retroviruses and Diseases They Cause.* Princeton: Excerpta Medica, 1988. Pp. 11–16.

Luc Montagnier, Willy Rozenbaum, and Jean-Claude Gluckman, eds. *SIDA et Infection par VIH.* Paris: Flammarion, 1989.

JAMES BLACK

James Black. "Pharmacology: Analysis and Exploration." *British Medical Journal* 293 (1986), 252–255.

James Black. "Basic Drug Research in Universities and Industry." *British Journal of Clinical Pharmacology* 22, Supplement 1, (1986), 5S–7S.

James Black. "Should We Be Concerned About the State of Hormone Receptor Classification?" In J.W. Black, D.H. Jenkinson, and V.P. Gerskowitch, eds. *Perspectives on Receptor Classification.* New York: Alan R. Liss, 1987. Pp. 11–15.

James Black. "Drugs from Emasculated Hormones: The Principle of Syntopic Antagonism." *Science* 245, 4917 (1989), 486–493.

THOMAS ADEOYE LAMBO

T. Adeoye Lambo. "Further Neuropsychiatric Observations in Nigeria." *British Medical Journal* 5214 (December 10, 1960), 1696–1704.

T. Adeoye Lambo. "Malignant Anxiety: A Syndrome Associated with Criminal Conduct in Africans." *The Journal of Mental Science* 108 (May 1962), 256–264.

T. Adeoye Lambo. "The African Mind in Contemporary Conflict." *WHO Chronicle* 25, 8 (1971), 343–353.

T. Adeoye Lambo. "Psychotherapy in Africa." *Psychotherapy and Psychosomatics* 24, 4–6 (1974), 311–326.

Alexander H. Leighton, T. Adeoye Lambo, Charles C. Hughes, Dorothea C. Leighton, Jane M. Murphy, and David B. Macklin. *Psychiatric Disorder Among the Yoruba.* Ithaca: Cornell University Press, 1963.

ETIENNE-EMILE BAULIEU

Etienne-Emile Baulieu and Sheldon Segal, eds. *The Antiprogestin Steroid RU 486 and Human Fertility Control.* New York: Plenum, 1985.

Etienne-Emile Baulieu. "Contragestion and Other Clinical Applications of RU 486, an Antiprogesterone at the Receptor." *Science* 245, 4924 (1989), 1351–1357.

Etienne-Emile Baulieu. "RU-486 as an Antiprogesterone Steroid." *JAMA* 262, 13 (1989), 1808–1814.

Etienne-Emile Baulieu. *Génération Pilule.* Paris: Odile Jacob, 1990.

Etienne-Emile Baulieu, with Mort Rosenblum. *The "Abortion Pill" RU-486: A Woman's Choice.* New York: Simon & Schuster, 1991.

RICHARD DAWKINS

Richard Dawkins. *The Extended Phenotype.* Oxford: W.H. Freeman, 1982.

Richard Dawkins. *The Blind Watchmaker.* Harlow: Longman, 1986.

Richard Dawkins. "The Evolution of Evolvability." In Chris Langton, ed. *Artificial Life.* Reading, MA: Addison-Wesley, 1989. Pp. 201–220.

Richard Dawkins. *The Selfish Gene.* New Edition. Oxford: Oxford University Press, 1989.

FAROUK EL-BAZ

Farouk El-Baz. "Desert Builders Knew a Good Thing When They Saw It." *Smithsonian* (April 1981), 116–122.

Farouk El-Baz. "Origin and Evolution of the Desert." *Interdisciplinary Science Reviews* 13, 4 (1988), 331–347.

Farouk El-Baz. "Finding a Pharaoh's Funeral Bark." *National Geographic* (April 1988), 512–533.

Farouk El-Baz. "Remote Sensing at an Archaeological Site in Egypt." *American Scientist* 77, 1 (1989), 60–66.

BERT SAKMANN

Erwin Neher and Bert Sakmann. "The Patch Clamp Technique." *Scientific American* (March 1992), 44–51.

Bert Sakmann and Erwin Neher. *Single-Channel Recording.* New York: Plenum, 1983.

Bert Sakmann. "Ion Channels and Signal Transmission in the Nervous System." *Les Cahiers de la Fondation Louis Jeantet de Médecine* 3 (1988).

Bert Sakmann. "Elementary Steps in Synaptic Transmission Revealed by Currents through Single Ion Channels." *Neuron* 8, 4 (April 1992), 613–629.

JONATHAN MANN

Jonathan Mann. "AIDS in Africa." *New Scientist* (26 March 1987), 40–43.

Jonathan Mann, James Chin, Peter Piot, and Thomas Quinn. "The International Epidemiology of AIDS. *Scientific American* (October 1988), 82–89.

Jonathan Mann. "AIDS in the 1990s: A Global Analysis." *Journal of the Royal Society of Health* 112, 3 (1992), 143–148.

Jonathan Mann, Daniel J.M. Tarantola, and Thomas W. Netter, eds. *AIDS in the World*. Cambridge: Harvard University Press, 1992.

Jaime Sepulveda, Harvey Fineberg, and Jonathan Mann, eds. *AIDS: Prevention Through Education: A World View*. New York: Oxford University Press, 1992.

NORMAN PACKARD

Mark Bedau and Norman Packard. "Measurement of Evolutionary Activity, Teleology, and Life." In Christopher G. Langton, et al., eds. *Artificial Life II*. Redwood City, CA: Addison-Wesley, 1991. Pp. 431–461.

James Crutchfield, Doyne Farmer, Norman Packard, and Rob Shaw. "Chaos." *Scientific American* (December 1986), 38–49.

Doyne Farmer and Norman Packard. "Evolution, Games, and Learning: Models for Adaptation in Machines and Nature." *Physica D* 22, 1–3 (1986), 7–12.

Doyne Farmer, Stuart Kauffman, Norman Packard, and Alan Perelson. "Adaptive Dynamic Networks as Models for the Immune-System and Autocatalytic Sets." *Annals of the New York Academy of Sciences* 50, 4 (1987), 118–131.

Norman Packard. "A Genetic Learning Algorithm for the Analysis of Complex Data." *Complex Systems* 4, 5 (1990), 543–572.

MARY-CLAIRE KING

Mary-Claire King and A.C. Wilson. "Evolution at Two Levels in Humans and Chimpanzees." *Science* 188, 4184 (1975), 107–116.

Mary-Claire King. "Genetic Analysis of Cancer in Families." *Cancer Surveys* 9, 3 (1990), 417–435.

Mary-Claire King. "Localization of the Early-Onset Breast Cancer Gene." *Hospital Practice* 26, 10 (1991), 121–126.

Mary-Claire King. "An Application of DNA Sequencing to a Human Rights Problem." *Molecular Genetic Medicine* 1 (1991), 117–131.

"Breast Cancer Research: A Special Report." *Science* 259, 5059 (1993), 616–632.

Photograph Credits

✦ ✦ ✦

p. 6, Sarah Hrdy, by Tom Zimberoff

p. 26, Luc Montagnier, by Henner Prefi

p. 48, James Black, by James Merrell

p. 66, Thomas Adeoye Lambo, by Mike Mitchell

p. 88, Etienne-Emile Baulieu, by Nick Clark

p. 108, Richard Dawkins, by Stephen Hyde

p. 130, Farouk El-Baz, by Ruvén Afanador

p. 150, Bert Sakmann, by Jean Mohr

p. 174, Jonathan Mann, by Paul Schirnhofer

p. 198, Norman Packard, by Michael Dismatsek

p. 218, Mary-Claire King, by Jane Scherr